Raspberry Pi Cookbook

Simon Monk

Beijing · Cambridge · Farnham · Köln · Sebastopol · Tokyo

Raspberry Pi Cookbook

by Simon Monk

Copyright © 2014 Simon Monk. All rights reserved.

Printed in the United States of America.

Published by O'Reilly Media, Inc., 1005 Gravenstein Highway North, Sebastopol, CA 95472.

O'Reilly books may be purchased for educational, business, or sales promotional use. Online editions are also available for most titles (*http://my.safaribooksonline.com*). For more information, contact our corporate/institutional sales department: 800-998-9938 or *corporate@oreilly.com*.

Editor: Rachel Roumeliotis	**Indexer:** Judy McConville
Production Editor: Melanie Yarbrough	**Cover Designer:** Karen Montgomery
Copyeditor: Nancy Reinhardt	**Interior Designer:** David Futato
Proofreader: Rachel Monaghan	**Illustrator:** Rebecca Demarest

December 2013: First Edition

Revision History for the First Edition:

2013-12-06: First release

See *http://oreilly.com/catalog/errata.csp?isbn=9781449365226* for release details.

Nutshell Handbook, the Nutshell Handbook logo, and the O'Reilly logo are registered trademarks of O'Reilly Media, Inc. *Raspberry Pi Cookbook*, the image of a Eurasian sparrowhawk, and related trade dress are trademarks of O'Reilly Media, Inc.

Many of the designations used by manufacturers and sellers to distinguish their products are claimed as trademarks. Where those designations appear in this book, and O'Reilly Media, Inc., was aware of a trademark claim, the designations have been printed in caps or initial caps.

ISBN: 978-1-449-36522-6

[LSI]

Table of Contents

Preface

Since its launch in 2011, the Raspberry Pi has found a role both as a very low-cost Linux-based computer and as a platform for embedded computing. It has proven popular with educators and hobbyists alike, with over 2 million units sold since its release.

In this book, you will find a wide range of recipes using the Raspberry Pi, including recipes for getting started and setting up your Pi; recipes for using the Python programming language; and a large number of recipes about using the Raspberry Pi with sensors, displays, motors, and so on. The book also includes a chapter on using the Raspberry Pi with Arduino boards.

This book is designed in such a way that you can read it linearly, as you would a regular book, or access recipes at random. You can search the table of contents or index for the recipe that you want and then jump right to it. If the recipe requires you to know about other things, then it will refer you to other recipes, rather like a cookbook might refer you to base sauces before showing you how to cook something fancier.

The world of Raspberry Pi is one that moves quickly. With a large active community, new interface boards and software libraries are being developed all the time. So, besides many examples that use specific interface boards or pieces of software, the book also covers basic principles so that you can have a better understanding of how to use new technologies that come along as the Raspberry Pi ecosystem develops.

As you would expect, there is a large body of code (mostly Python programs) that accompanies the book. These programs are all open source and available on GitHub. You'll find a link to them at the *Raspberry Pi Cookbook* website (*http://www.raspberry picookbook.com*).

For most of the software-based recipes, all you need is a Raspberry Pi. I recommend a Raspberry Pi model B. For recipes that involve making your own hardware to interface with the Raspberry Pi, I have tried to make good use of ready-made modules, as well as solderless breadboard and jumper wires to avoid the need for soldering.

For those wishing to make breadboard-based projects more durable, I suggest using protoyping boards with the same layout as a half-sized breadboard, such as those sold by Adafruit, so that the design can easily be transferred to a soldered solution.

Conventions Used in This Book

The following typographical conventions are used in this book:

Italic
> Indicates new terms, URLs, email addresses, filenames, and file extensions.

`Constant width`
> Used for program listings, as well as within paragraphs to refer to program elements such as variable or function names, databases, data types, environment variables, statements, and keywords.

`Constant width bold`
> Shows commands or other text that should be typed literally by the user.

`Constant width italic`
> Shows text that should be replaced with user-supplied values or by values determined by context.

 This icon signifies a tip, suggestion, or general note.

 This icon indicates a warning or caution.

 This icon points you to the related video for that section.

Using Code Examples

Supplemental material (code examples, exercises, etc.) is available for download at *http://www.raspberrypicookbook.com*.

This book is here to help you get your job done. In general, if example code is offered with this book, you may use it in your programs and documentation. You do not need to contact us for permission unless you're reproducing a significant portion of the code. For example, writing a program that uses several chunks of code from this book does

not require permission. Selling or distributing a CD-ROM of examples from O'Reilly books does require permission. Answering a question by citing this book and quoting example code does not require permission. Incorporating a significant amount of example code from this book into your product's documentation does require permission.

We appreciate, but do not require, attribution. An attribution usually includes the title, author, publisher, and ISBN. For example: "*Raspberry Pi Cookbook* by Simon Monk (O'Reilly). Copyright 2014 Simon Monk, 978-1-449-36522-6."

If you feel your use of code examples falls outside fair use or the permission given above, feel free to contact us at *permissions@oreilly.com*.

Safari® Books Online

Safari Books Online is an on-demand digital library that delivers expert content in both book and video form from the world's leading authors in technology and business.

Technology professionals, software developers, web designers, and business and creative professionals use Safari Books Online as their primary resource for research, problem solving, learning, and certification training.

Safari Books Online offers a range of product mixes and pricing programs for organizations, government agencies, and individuals. Subscribers have access to thousands of books, training videos, and prepublication manuscripts in one fully searchable database from publishers like O'Reilly Media, Prentice Hall Professional, Addison-Wesley Professional, Microsoft Press, Sams, Que, Peachpit Press, Focal Press, Cisco Press, John Wiley & Sons, Syngress, Morgan Kaufmann, IBM Redbooks, Packt, Adobe Press, FT Press, Apress, Manning, New Riders, McGraw-Hill, Jones & Bartlett, Course Technology, and dozens more. For more information about Safari Books Online, please visit us online.

How to Contact Us

Please address comments and questions concerning this book to the publisher:

O'Reilly Media, Inc.
1005 Gravenstein Highway North
Sebastopol, CA 95472
800-998-9938 (in the United States or Canada)
707-829-0515 (international or local)
707-829-0104 (fax)

We have a web page for this book, where we list errata, examples, and any additional information. You can access this page at *http://oreil.ly/raspberry-pi-cb*.

To comment or ask technical questions about this book, send email to *bookques tions@oreilly.com.*

For more information about our books, courses, conferences, and news, see our website at *http://www.oreilly.com.*

Find us on Facebook: *http://facebook.com/oreilly*

Follow us on Twitter: *http://twitter.com/oreillymedia*

Watch us on YouTube: *http://www.youtube.com/oreillymedia*

Acknowledgments

As always, I thank Linda for her patience and support.

I also thank the technical reviewers Duncan Amos, Chaim Krause, and Steve Suehring; their comments were very helpful.

Rachel Roumeliotis has done a great job as editor on this project. Her pragmatic approach is very easy to work with and has made this a very interesting project to work on.

Thanks also to all the O'Reilly team, especially those I met at the Cambridge office, who were very welcoming when I visited, and of course Nan Reinhardt for her diligent copyediting.

Setup and Management

1.0. Introduction

When you buy a Raspberry Pi, you are essentially buying an assembled printed circuit board. It does not even include a power supply or operating system.

The recipes in this chapter are concerned with getting your Raspberry Pi set up and ready for use.

Because the Raspberry Pi just uses standard USB keyboards and mice, most of the setup is pretty straightforward, so we will concentrate on those tasks that are specific to the Raspberry Pi.

1.1. Selecting a Model of Raspberry Pi

Problem

There are two models of Raspberry Pi, A and B. You are not sure which to use.

Solution

If you want a Raspberry Pi for general use, then you should buy a model B, revision 2 (the latest). With twice as much memory, it will cope with most tasks much better than the model A.

If, on the other hand, you are embedding a Raspberry Pi in a project for a single purpose, then using a model A and saving a few dollars may well be an option.

Discussion

Figure 1-1 shows model A and model B side by side.

Figure 1-1. Raspberry Pi model A (left) and model B (right)

As you can see from Figure 1-1, both use the same circuit board, but model A has a single USB socket and no RJ45 Ethernet socket. You can also see the bare solder pads behind the USB socket where the Ethernet controller chip should be.

The differences between the models are summarized in Table 1-1. This table also includes the original model B, revision 1 board, which was soon updated with revision 2. You can tell if you have a revision 1 board because it has a black audio socket rather than the blue socket of the revision 2 board.

Table 1-1. Raspberry Pi models

Model	RAM	USB sockets	Ethernet port
A	256 MB	1	no
B rev2	512 MB	2	yes
B rev1	256 MB	2	yes

The lack of a network interface on the model A is not a problem because all the Raspberry Pi models will accept a USB WiFi adapter (Recipe 2.5). Since the adapter would use the only USB socket, you would also need a USB hub to increase the number of USB sockets. However, with a model B Raspberry Pi, you can put a USB WiFi adapter in one USB socket and a wireless keyboard/mouse dongle in the other.

See Also

For more information on the Raspberry Pi models, see *http://en.wikipedia.org/wiki/Raspberry_PI*.

1.2. Enclosing a Raspberry Pi

Problem

You need an enclosure for your Raspberry Pi.

Solution

The Raspberry Pi does not come with an enclosure unless you buy one as part of a kit. This makes it a little vulnerable, as there are bare connections on the underside of the circuit board that could easily short if the Raspberry Pi is placed on something metal.

It is a good idea to buy some protection for your Raspberry Pi in the form of a case.

Discussion

There is a vast array of case styles to choose from, including:

- Simple, two-part, click-together plastic boxes
- VESA mountable boxes (for attaching to the back of a monitor or TV)
- Lego-style boxes
- 3D printed box designs
- Laser-cut, snap-together acrylic designs

The case you buy is very much a matter of personal taste. However, some of the things you need to consider are:

- Do you need to have access to the GPIO connector? This is important if you plan to attach external electronics to your Raspberry Pi.
- Is the case well ventilated? This is important if you plan to overclock your Raspberry Pi (Recipe 1.14) or run it hard, playing videos or games, as these will all generate more heat.

See Also

Adafruit has a nice range of Raspberry Pi enclosures (*http://bit.ly/1aDT3qm*).

You will also find many styles of cases at other Raspberry Pi suppliers and on eBay.

1.3. Selecting a Power Supply

Problem

You need to select a power supply for your Raspberry Pi.

Solution

The basic electrical specification for a power supply suitable for a Raspberry Pi is that it supplies a regulated 5V DC (direct current) and can supply a current of up to 700mA. It must also have a micro USB plug on the end of the lead.

If you buy your power supply from the same place as the Raspberry Pi, then the seller should be able to tell you if it will work with the Raspberry Pi.

If you are going to be using a WiFi dongle, or other USB peripherals that use significant amounts of power, then I would get a power supply capable of 1.5A or even 2A. Also beware of very low-cost power supplies that may not provide an accurate or reliable 5V.

Discussion

The power supply and connector are actually the same as those found in many smartphone chargers. If they are terminated in a micro USB plug, then they are almost certainly 5V (but check). The only question, then, is if they can supply enough current.

If they can't, then a few bad things can happen:

- They may get hot and be a potential fire risk.
- They may just fail.
- At times of high load (say, when the Pi is using a WiFi dongle), the voltage may dip, and the Raspberry Pi may reset itself.

So, look for a power supply that says it can supply 700 mA or more. If it specifies a number of watts (W) rather than mA, then divide the number of watts by 5 to get the mA figure. So, a 5V 10W power supply can supply 2A (2,000 mA).

Using a power supply with, say, a maximum current of 2A will not use any more electricity than a 700mA power supply. The Raspberry Pi will just take as much current as it needs.

In Figure 1-2, I have measured the current taken by a Raspberry Pi model B, revision 2, as it starts up, with and without a WiFi dongle and HDMI video connection using the Raspian Wheezy distribution.

You can see that actually the current rarely gets above 500mA. However, the processor isn't really doing very much here. Were you to start playing HD video, the current would

Figure 1-2. Raspberry Pi current consumption during booting

increase considerably. When it comes to power supplies, it's usually better to have something in reserve.

See Also

You can buy a module that will turn off the power when the Raspberry Pi shuts down at *http://www.pi-supply.com/*.

1.4. Selecting an Operating System Distribution

Problem

There are a number of different Raspberry Pi distributions. You are not sure which one to use.

Solution

The answer to this question depends on what you intend to do with your Raspberry Pi.

If you plan to build hardware projects with the Raspberry Pi, then use Raspbian or Occidentalis (from Adafruit). Of these, Raspbian is the official and most commonly used distribution for Raspberry Pi, but Occidentalis requires less setup to start hacking hardware.

If you plan to use your Raspberry Pi as a media center, there are a number of distributions specifically for that purpose (see Recipe 4.1).

In this book, we use the Raspbian distribution almost exclusively, although most of the recipes will work with any Debian-based distribution.

Discussion

SD cards—especially a 4 GB SD card, which is recommended for most distributions— are not expensive, so get a few and try out a few distributions. If you do this, it is a good idea to keep your own files on a USB flash drive so that you don't have to keep copying them onto each SD card.

Note that if you are using one of the upcoming recipes, to write your own SD card, then you need to either have a computer that has an SD card slot (many laptops do) or buy an inexpensive USB SD card reader.

See Also

The official list of Raspberry Pi distributions (*http://www.raspberrypi.org/downloads*)

See Occidentalis (*http://bit.ly/adafruit-v2*)

1.5. Writing an SD Card with NOOBS

Problem

You want to write an SD card using NOOBS (New Out of the Box Software).

Solution

Download the NOOBS archive file from *http://www.raspberrypi.org/downloads*, extract it, and place it on an SD card.

 NOOBS downloads the software it needs, so if you are using a Raspberry Pi model A without a network connection, then you may find it best to use one of recipes that installs the whole operating system (Recipes 1.6, 1.7, or 1.8).

Once you have downloaded the NOOBS archive file, extract it and copy the folder contents onto the SD card. Note that if the archive extracts to a folder called *NOOBS_v1_2_1* or similar, it is the contents of the folder that should be copied to the root of the SD card, not the folder itself.

Put the SD card containing the extracted NOOBS files into your Raspberry Pi and then power up your Raspberry Pi. When it boots, the window shown in Figure 1-3 will appear. From this screen, you can select which distribution you want to install. Raspbian will be selected as the default, and this is the best choice to start off with.

![NOOBS first screen showing NOOBS v1.2.1 - Built: Jun 26 2013 with Install OS, Edit config (e), Online help (h), Exit (Esc) menu options, and a list of distributions: Archlinux, OpenELEC, Pidora, RISC OS, RaspBMC, and Raspbian [RECOMMENDED]]

Figure 1-3. NOOBS first screen

You will get a warning message that the SD card will be overwritten (which is fine) and then as the distribution is installed onto the SD card, you will see a progress screen accompanied by helpful information about the distribution (Figure 1-4).

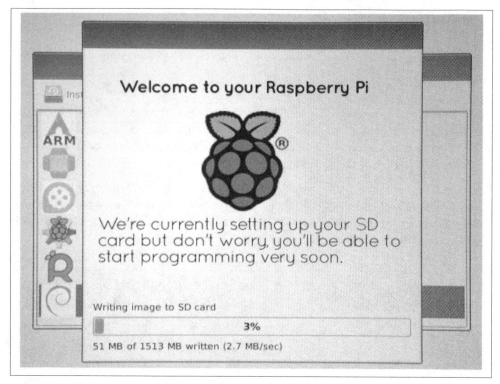

Figure 1-4. NOOBS overwriting the SD card

Once the file copying is complete, you will get the message *Image applied successfully*. When you hit return, the Raspberry Pi will reboot and then *raspi_config* will automatically run so that you can configure the new installation (Recipe 1.12).

Discussion

For NOOBS to install correctly onto an SD crd, the card must be formatted as FAT. Most SD cards are supplied already formatted in FAT. If you are reusing an old card and need to format it as FAT, then use your operating system's tool for formatting removeable media.

Alternatively, the Raspberry Pi foundation recommends the SD Association's formatting tool (*https://www.sdcard.org/downloads/formatter_4/*), which is available for Mac and Windows.

See Also

See Recipes 1.6, 1.7, or 1.8 to write a full operating system image to an SD card.

1.6. Writing an SD Card Manually (Mac)

Problem

The Raspberry Pi is not generally supplied with an operating system. You want to use a Mac to write an operating system distribution onto an SD card.

Solution

Download an *iso* disk image for the distribution that you have selected (see Recipe 1.4). Then use a utility script to write it onto an SD card.

There are a number of different utility programs that have been written to help you create an SD card. A popular one is available on GitHub (*http://bit.ly/viljoen-installer*).

1. Download it and extract the folder contained in the zip file to some convenient location.

2. Copy the *iso* image file you downloaded into this folder. Let's assume it is something like *Occidentalis_v02.iso*.

3. Eject any external hard disks or USB flash drives and insert the SD card.

4. Open a Terminal session, change directory (cd command) to the installer folder, and run this command:

```
sudo ./install Occidentalis_v02.iso
```

If you are using a different *iso* image file, then use that filename instead. This command launches the SD installer, and you'll see something like Figure 1-5.

Figure 1-5. Writing an SD card (Mac)

You now need to enter a number corresponding to the SD card. This is why it's a good idea to eject all other removable storage media, because if you pick the wrong option, the utility will trash whatever is on that drive. So, before continuing, make sure that you know which number corresponds to your SD card.

After you have made the selection, if all is well, writing will commence. This usually takes a few minutes.

Discussion

You can also buy preformatted SD cards with a distribution already installed on them. However, these are sometimes not the latest version, so it's a good idea to know how to write your own SD cards.

When selecting an SD card to write to, always select one that is at least 4 GB.

See Also

See also Recipe 1.5 for how to write an SD card image by using NOOBS.

All the (*)Ware (*http://bit.ly/sd-setup*) has an alternative utility for writing an SD card from a Mac.

More details about writing an SD card are available at *http://bit.ly/1cqhXym*.

1.7. Writing an SD Card Manually (Windows)

Problem

The Raspberry Pi is not generally supplied with an operating system. You want to use a Windows PC to write an operating system distribution onto an SD card.

Solution

Download an *iso* disk image for the distribution that you have selected (see Recipe 1.4). Then use the Fedora ARM Installer to put it onto an SD card.

Start by downloading the Fedora ARM Installer (*http://bit.ly/ISLPc4*).

Note that this will only work on Windows Vista and newer.

- Extract the folder contained in the zip file to some convenient location (say, the desktop).
- Eject any external hard disks or USB flash drives and insert the SD card.
- Run *fedora-arm-installer.exe* as a user who has administrator rights.

This tool has a nice user interface (Figure 1-6).

Figure 1-6. Writing an SD card (Windows)

Simply navigate to wherever you downloaded the *iso* file in the Source section, select the SD card from the Destination drop-down, and after checking that you have definitely selected the SD card and not some other disk drive, click Install.

Discussion

You can also buy preformatted SD cards with a distribution already installed on them. However, these are sometimes not the latest version, so it is a good idea to know how to write your own SD cards.

When selecting an SD card to write to always select one that is at least 4GB.

See Also

See also Recipe 1.5 for how to write an SD card image by using NOOBS.

More details and options for writing an SD card are available at *http://elinux.org/ RPi_Easy_SD_Card_Setup*.

1.8. Writing an SD Card Manually (Linux)

Problem

The Raspberry Pi is not generally supplied with an operating system. You want to use a Linux PC to write an operating system distribution onto an SD card.

Solution

Download an *iso* disk image for the distribution that you have selected (see Recipe 1.4). Then use the Ubuntu `ImageWriter` tool to put it onto an SD card. For other Linux distributions, consider using NOOBS (Recipe 1.5).

First, install the `ImageWriter` tool by opening a Terminal session and entering the following command:

```
$ sudo apt-get install usb-imagewriter
```

This tool has a nice user interface (Figure 1-7).

Figure 1-7. Writing an SD Card (Linux)

Navigate to wherever you downloaded the *iso* file in the Write Image section, select the SD card from the "to" drop-down, and after checking that you have definitely selected the SD card and not some other disk drive, click "Write to Device."

Discussion

You can also buy preformatted SD cards with a distribution already installed on them. However, these are sometimes not the latest version, so it is a good idea to know how to write your own SD cards.

When selecting an SD card to write to, always select one that is at least 4 GB.

See Also

See also Recipe 1.5 for how to write an SD card image by using NOOBS.

More details and options for writing an SD card are available at *http://elinux.org/RPi_Easy_SD_Card_Setup*.

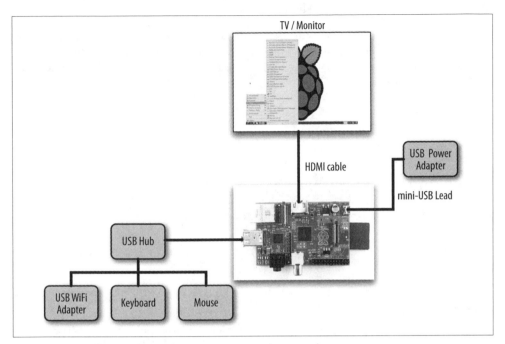

Figure 1-8. A typical Raspberry Pi system

1.9. Connecting the System

Problem

You have everything that you need for your Raspberry Pi, and you want to connect it all together.

Solution

Unless you are embedding your Raspberry Pi in a project or using it as a media center, you need to attach a keyboard, mouse, monitor, and probably a WiFi dongle. As this adds up to more than three connections, even with a Raspberry Pi model B, you'll need a USB hub to provide enough USB sockets.

Figure 1-8 shows a typical Raspberry Pi system.

Discussion

If you use a wireless keyboard/mouse combination with a single USB dongle, then the other free USB socket (assuming you have a Raspberry Pi model B) could be used for

a WiFi dongle. However, in all likelihood a USB hub is going to be useful when you want to attach an external USB disk or USB flash drive.

The Raspberry Pi is perfectly happy with pretty much any keyboard or mouse, wired or wireless. The exception to this is Bluetooth wireless keyboards and mice, which will not work with the Raspberry Pi.

See Also

The official Raspberry Pi Quick Start Guide (*http://bit.ly/1ju8usM*)

1.10. Connecting a DVI or VGA Monitor

Problem

Your monitor does not have an HDMI connector, but you want to use it with your Raspberry Pi.

Solution

Many people have been caught out by this problem. Fortunately, it is possible to buy adapters for monitors with a DVI or VGA input, but not HDMI connectors.

DVI adapters are the simplest and cheapest. They can be found for less than $5 if you search for "HDMI male to DVI female converter."

Discussion

Using VGA adapters is more complex, as they require some electronics to convert the signal from digital to dialog, so beware of leads that do not contain these. The official converter is called *Pi-View* and is available wherever the Raspberry Pi is sold. Pi-View has the advantage of having been tested and found to work with Raspberry Pi. You may find cheaper alternatives on the Internet, but often these won't work.

See Also

elinux has tips on what to look for in an converter (*http://bit.ly/rpi-peripherals*).

1.11. Using a Composite Video Monitor/TV

Problem

The text on your low-resolution composite monitor is illegible. You need to adjust the resolution of the Raspberry Pi for a small screen.

Solution

The Raspberry Pi has two types of video output, HDMI and composite video. Of these, the HDMI is much better quality. So if you're intending to use a composite video as your main screen, you may want to think again.

If you are using such a screen—say, because you need a really small screen—then you need to make a few adjustments to fit the video output to the screen. You need to make some changes to the file */boot/config.txt*. You can edit it on the Raspberry Pi by issuing the following command in a Terminal session:

```
$ sudo nano /boot/config.txt
```

If the text is too small to read and you do not have an HDMI monitor, then you can also edit the file by removing the SD card from the Raspberry Pi and inserting it into your computer. The file will then be in the top-level directory on the SD card, so you can use a text editor on your PC to modify it.

You need to know the resolution of your screen. For a lot of small screens, this will be 320 by 240 pixels. Find the two lines in the file that read:

```
#framebuffer_width=1280
#framebuffer_height=720
```

Remove the # from the front of each line and change the two numbers to the width and height of your screen. In the following example, these lines have been modified to be 320 by 240:

```
framebuffer_width=320
framebuffer_height=240
```

Save the file and restart your Raspberry Pi. You should find that everything has become a lot easier to read. You will probably also find that there is a big thick border around the screen. To adjust this, see Recipe 1.13.

Discussion

There are many low-cost CCTV monitors that can make a great companion for the Raspberry Pi when you're making something like a retro games console (Recipe 4.6). However, these monitors are often very low resolution.

See Also

For another tutorial on using composite monitors, see this Adafruit's tutorial (*http://bit.ly/adafruit-learning*).

Also, see Recipe 1.13 and Figure 1-10 to adjust your picture when you're using the HDMI video output.

```
Raspi-config

        info              Information about this tool
        expand_rootfs     Expand root partition to fill SD card
        overscan          Change overscan
        configure_keyboard Set keyboard layout
        change_pass       Change password for 'pi' user
        change_locale     Set locale
        change_timezone   Set timezone
        memory_split      Change memory split
        ssh               Enable or disable ssh server
        boot_behaviour    Start desktop on boot?
        update            Try to upgrade raspi-config

           <Select>                        <Finish>
```

Figure 1-9. Adjusting the root partition

1.12. Using All the Storage on the SD Card

Problem

When an operating system is written to a Raspberry Pi's SD card, the partition size is fixed by the image you install so you cannot use the whole of the SD card. This leaves little room for your own files.

Solution

To fix this, you need to run the raspi-config tool. This program automatically runs the first time that you boot your Raspberry Pi with a new SD card. You can also start it any time you need to adjust the configuration by opening a Terminal session and issuing the command:

```
$ sudo raspi-config
```

Scroll down to the option expand_rootfs, and then use the cursor keys to press the Select button (Figure 1-9).

You will see a confirmation message, and then your Raspberry Pi will need to restart for the change to take effect.

You should now be able to use the full capacity of the SD card.

Discussion

Adjusting the root partition is definitely worth doing, especially for storing temporary files, but keeping your documents and personal files on a USB flash drive avoids the need to transfer files when you update your operating system.

See Also

You can find much more information about the `raspi-config` tool at *http://elinux.org/RPi_raspi-config*.

1.13. Adjusting the Picture Size on your Monitor

Problem

When you first connect a Raspberry Pi to a monitor, you may find that some of the text cannot be read because it extends off the screen, or the picture isn't using all the space available on the screen.

Solution

If your text extends off the screen, use the `raspi-config` tool to turn `overscan` off.

To do this, run `raspi-config` by opening a Terminal session and issuing the command:

```
$ sudo raspi-config
```

Then use the cursor keys to scroll down to the option overscan, and turn overscan off (Figure 1-10).

If your problem is that there is a large black border around the picture, then you can reduce this (and possibly eliminate it entirely) by editing the file /boot/config.txt using the command:

```
$ sudo nano /boot/config.txt
```

Look for the section dealing with overscan. The four lines you need to change are shown in the middle of Figure 1-11.

For the lines to take effect, you first need to uncomment them by removing the # character from the start of each line.

Then, using trial and error, change the settings until the screen fills as much of the monitor as possible. Note that the four numbers should be negative. Try setting them all to −20 to start with.

Figure 1-10. Turning off overscan

Figure 1-11. Adjusting overscan

Discussion

Having to repeatedly restart the Raspberry Pi to see the effects of the changes in resolution is a little tedious. Fortunately, you will only have to do this procedure once. Many monitors and TVs work just fine without any adjustments being necessary.

See Also

You can find much more information about the `raspi-config` tool at *http://elinux.org/RPi_raspi-config*.

1.14. Maximizing Performance

Problem

Your Raspberry Pi seems to be very slow, so you want to overclock it to make it run faster.

Solution

You can increase the clock frequency of a Raspberry Pi to make it run a little faster. This will make it use a bit more power and run a little hotter (see the Discussion next).

The method of overclocking described here is called *dynamic overclocking* because it automatically monitors the temperature of the Raspberry Pi and drops the clock speed back down if things start to get too hot.

To make your Pi overclock, run the `raspi_config` utility by issuing the following command in a Terminal:

```
$ sudo raspi-config
```

Select the overclock option in the menu, and you are presented with the options in Figure 1-12.

Select an option. If you find that your Raspberry Pi starts to become unstable and hangs unexpectedly, then you may need to choose a more conservative option or turn overclocking off by setting it back to None.

Discussion

The performance improvements from overclocking can be quite dramatic. To measure these, I used a Raspberry Pi model B, revision 2, without a case at an ambient room temperature of 15 degrees C.

The test program was the following Python script. This just hammers the processor and is not really representative of the other things that go on in a computer, such as writing to the SD card, graphics, and so on. But it does give a good indication of raw CPU performance if you want to test the effect of overclocking on your Raspberry Pi.

```
import time

def factorial(n):
    if n == 0:
```

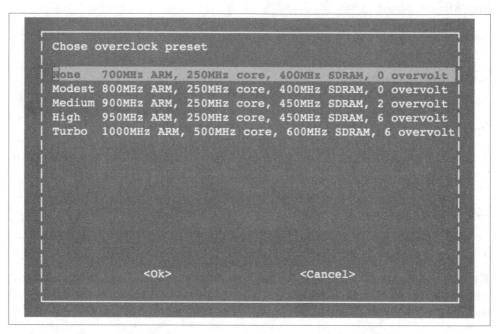

Figure 1-12. Overclocking options

```
    return 1
  else:
    return n * factorial(n-1)

before_time = time.clock()
for i in range(1, 10000):
  factorial(200)
after_time = time.clock()

print(after_time - before_time)
```

Check out the results of the test in Table 1-2.

Table 1-2. Overclocking

	Speed test	Current	Temperature (degrees C)
700 MHz	15.8 seconds	360 mA	27
1 GHz	10.5 seconds	420 mA	30

As you can see, the performance has increased by 33%, but at a cost of drawing more current and a slightly higher temperature.

You can now buy self-adhesive heatsinks that fit onto the big chip at the center of the Raspberry Pi to help keep it cool. Some of these do not actually connect the heatsink to

the chip with a thermally conductive compound and are therefore mostly for show. A well-ventilated enclosure (Recipe 1.2) may well be more effective. There have also been some efforts to add water-cooling to the Raspberry Pi. Frankly, this is just silly.

See Also

You can find much more information about the `raspi-config` tool at *http://elinux.org/ RPi_raspi-config*.

1.15. Changing Your Password

Problem

By default, the password for a Raspberry Pi will be *raspberry*. You want to change this.

Solution

You can use the `raspi-config` tool to change your password. Run the `raspi_config` utility by issuing the following command in a Terminal (see Recipe 3.2):

```
$ sudo raspi-config
```

Then select the change_pass option in the menu and follow the prompts shown in Figure 1-13.

Changing your password is one occasion where you do not have to restart your Raspberry Pi for the changes to take effect.

Discussion

You can also change the password from a Terminal session simply by using the `passwd` command as follows:

```
$ sudo passwd
Changing password for pi.
(current) UNIX password:
Enter new UNIX password:
Retype new UNIX password:
passwd: password updated successfully
```

See Also

You can find much more information about the `raspi-config` tool at *http://elinux.org/ RPi_raspi-config*.

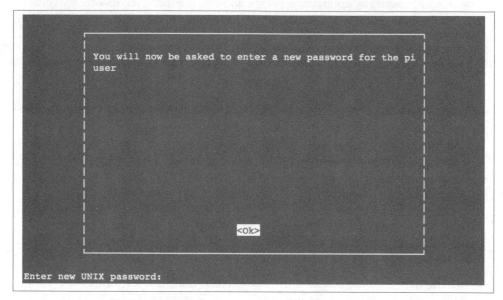

Figure 1-13. Changing the password

1.16. Setting the Pi to Boot Straight into a Windowing System

Problem

Every time you reboot your Raspberry Pi, you have to log in and then start the desktop manually. You want to make this automatic.

Solution

You can use the `raspi-config` tool to change the boot behavior so that the Raspberry Pi automatically logs you in and starts the desktop. Run the `raspi_config` utility by issuing the following command in a Terminal:

```
$ sudo raspi-config
```

Then select the boot_behaviour option in the menu and follow the prompts shown in Figure 1-14.

Figure 1-14. Automatic boot into a desktop

After you change the boot option, you are prompted to restart your Raspberry Pi for the changes to take effect.

Discussion

Clearly there are security implications in allowing the Raspberry Pi to log you in to the windowing environment automatically. But because the Raspberry Pi will generally be used as a personal computer, rather than being shared, the convenience usually outweighs any such disadvantages.

See Also

You can find much more information about the `raspi-config` tool at *http://elinux.org/RPi_raspi-config*.

1.17. Shutting Down Your Raspberry Pi

Problem

You want to shut down your Raspberry Pi.

Figure 1-15. Shutting down your Raspberry Pi

Solution

Click on the red Logout button in the bottom-right corner of the desktop. This will display a number of options (Figure 1-15).

Shutdown

Shuts down the Raspberry Pi. You will need to unplug the power and plug it in again to get the Raspberry Pi to boot up again.

Reboot

Reboots the Raspberry Pi.

Logout

Logs you out and displays a prompt to enter your login credentials so that you can log back in.

Cancel

Allows you to change your mind and continue using the Raspberry Pi.

You can also reboot from the command line by issuing the command:

```
sudo halt
```

You may have to do this after installing some software. When you do reboot, you see the message shown in Figure 1-16, which illustrates the multiuser nature of Linux and warns that all users connected to the Pi.

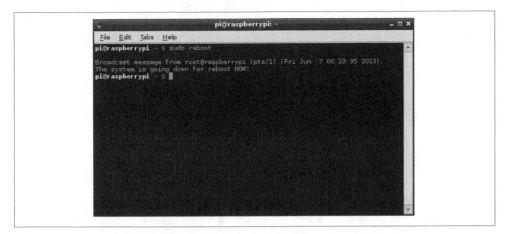

Figure 1-16. Shutting down your Raspberry Pi from the Terminal

Discussion

Unlike shutting down most computers, shutting down a Rasperry Pi does not actually turn off the power. It goes into a low-power mode—and it is a pretty low-power device anyway (but the Raspberry Pi hardware has no control over its power supply).

See Also

You can buy a module that will turn off the power when the Raspberry Pi shuts down here: *http://www.pi-supply.com/*.

1.18. Installing the Raspberry Pi Camera Module

Problem

You want to use the Raspberry Pi camera module (see "Modules" on page 381).

Solution

The Raspberry Pi camera module (Figure 1-17) is attached to a Raspberry Pi by a ribbon cable.

Figure 1-17. The Raspberry Pi camera module

This cable attaches to a special connector just behind the Ethernet socket. To fit it, pull up the levers on either side of the connector, so that they unlock, and then press the cable into the slot with the connector pads of the cable facing away from the Ethernet socket. Press the two levers of the connector back down to lock the cable in place (Figure 1-18).

Figure 1-18. Fitting a Raspberry Pi camera module

 The camera module packaging states that it is sensitive to static. So, before handling it, ground yourself by touching something grounded like the metal case of a PC.

The camera module requires some software configuration. The easiest way to configure it is to use `raspi_config` (Recipe 1.12). To run `raspi-config`, enter the following command into a Terminal session.

```
$ sudo raspi-config
```

You will then see a list of options like those shown in Figure 1-19.

```
● ● ●                    1234000000730 — pi@raspberrypi: ~ — ssh — 81×24

| Raspi-config
|
|          info                Information about this tool
|          expand_rootfs       Expand root partition to fill SD card
|          overscan            Change overscan
|          configure_keyboard  Set keyboard layout
|          change_pass         Change password for 'pi' user
|          change_locale       Set locale
|          change_timezone     Set timezone
|          memory_split        Change memory split
|          overclock           Configure overclocking
|          ssh                 Enable or disable ssh server
|          boot_behaviour      Start desktop on boot?
|          update              Try to upgrade raspi-config
|
|
|                 <Select>                           <Finish>
|
```

Figure 1-19. The raspi-config configuration tool

If the option Camera is not in the list, then you need to update the operating system by running the following commands in a Terminal session (see Recipe 3.2):

```
$ sudo apt-get update
$ sudo apt-get upgrade
```

You need an Internet connection to do this. The second of these commands may take a few minutes to complete. When they are complete, reboot your Raspberry Pi (Recipe 1.17).

Now, when you run `raspi-config` you will see the option to enable the camera (Figure 1-20).

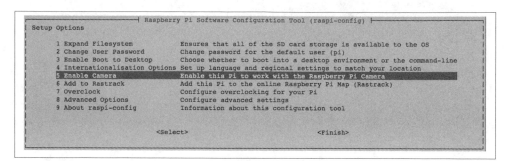

```
┌─────────┤ Raspberry Pi Software Configuration Tool (raspi-config) ├─────────┐
│ Setup Options                                                                │
│                                                                              │
│        1 Expand Filesystem          Ensures that all of the SD card storage is available to the OS
│        2 Change User Password       Change password for the default user (pi)
│        3 Enable Boot to Desktop     Choose whether to boot into a desktop environment or the command-line
│        4 Internationalisation Options Set up language and regional settings to match your location
│        5 Enable Camera              Enable this Pi to work with the Raspberry Pi Camera
│        6 Add to Rastrack            Add this Pi to the online Raspberry Pi Map (Rastrack)
│        7 Overclock                  Configure overclocking for your Pi
│        8 Advanced Options           Configure advanced settings
│        9 About raspi-config         Information about this configuration tool
│                                                                              │
│                                                                              │
│                     <Select>                              <Finish>           │
│                                                                              │
└──────────────────────────────────────────────────────────────────────────────┘
```

Figure 1-20. The updated raspi-config configuration tool

Two commands are available for capturing still images and videos: `raspiStill` and `raspivid`.

To capture a single still image, use the `raspiStill` command as shown here:

```
$ raspistill -o image1.jpg
```

A preview screen displays for about five seconds and then takes a photograph and stores it in the file *image1.jpg* in the current directory.

To capture video, use the command `raspivid`:

```
$ raspivid -o video.h264 -t 10000
```

The number on the end is the recording duration in milliseconds—in this case, 10 seconds.

Discussion

Both `raspstill` and `raspivid` have a large number of options. If you type either command without any parameters, help text displays options that are available.

The camera module is capable of high-resolution stills and video recording.

Here are some of the key features of the camera:

- 5-megapixel sensor
- Fixed focus f/2 lens
- Still resolution 1,920×1,080
- Video 1080p 30fps

An alternative to the camera module is to use a USB webcam (see Recipe 4.5).

See Also

The RaspiCam documentation (*http://bit.ly/cam-doc*) includes `raspstill` and `raspivid`.

Networking

2.0. Introduction

The Raspberry Pi is designed to be connected to the Internet. Its ability to communicate on the Internet is one of its key features and opens up all sorts of possible uses, including home automation, web serving, network monitoring, and so on.

The connection can be wired through an Ethernet cable (at least in the case of the model B), or the Pi can use a USB WiFi module to provide a network connection.

Having a connected Raspberry Pi also means that you can connect to it remotely from another computer. This is very useful in situations where the Raspberry Pi itself is inaccessible and does not have a keyboard, mouse, and monitor attached to it.

This chapter gives you recipes for connecting your Raspberry Pi to the Internet and controlling it remotely over a network.

2.1. Connecting to a Wired Network

Problem

You want to connect your Raspberry Pi to the Internet using a wired network connection.

Solution

First, if you have a Raspberry Pi model A, there is no RJ45 connector for Ethernet. In this case, your best option for Internet access is to use a wireless USB adaptor (see Recipe 2.5).

If you have a Raspberry Pi model B, plug an Ethernet patch cable into its RJ45 socket and then connect the other end to a spare socket on the back of your home hub/router (Figure 2-1).

Figure 2-1. Connecting Raspberry Pi to a home hub

The network LEDs on your Raspberry Pi should immediately start to flicker as the Raspberry Pi connects to your network.

Discussion

Raspbian and Occidentalis distributions (in fact, pretty much all distributions for the Pi) are preconfigured to connect to any network using DHCP (Dynamic Host Configuration Protocol). They will automatically be assigned an IP address as long as DHCP is enabled on your network.

If the network LEDs on your Raspberry Pi do not light up when you plug it into the home hub, check that you have not used the *Uplink* RJ45 socket on the hub or try a different cable.

If the LEDs blink, but you cannot connect to the Internet on your Raspberry Pi using a browser, check that DHCP is enabled on your network management console. Look for an option like that shown in Figure 2-2.

Figure 2-2. Enabling DHCP on your home hub

See Also

To connect to a wireless network, see Recipe 2.5.

2.2. Finding Out Your IP Address

Problem

You want to know the IP address of your Raspberry Pi so that you can communicate with it, whether connecting to it as a web server, exchanging files, or controlling it remotely with SSH (Recipe 2.7) or VNC (Recipe 2.8).

An IP address is a four-part number uniquely identifying a computer's network interface within a network. Each part is separated by a dot.

Solution

To find the IP address of your Raspberry Pi, you need to issue this command in a Terminal window:

```
$ sudo ifconfig
eth0      Link encap:Ethernet  HWaddr b8:27:eb:d5:f4:8f
          inet addr:192.168.1.16  Bcast:192.168.255.255  Mask:255.255.0.0
          UP BROADCAST RUNNING MULTICAST  MTU:1500  Metric:1
          RX packets:1114 errors:0 dropped:1 overruns:0 frame:0
          TX packets:1173 errors:0 dropped:0 overruns:0 carrier:0
          collisions:0 txqueuelen:1000
          RX bytes:76957 (75.1 KiB)  TX bytes:479753 (468.5 KiB)

lo        Link encap:Local Loopback
          inet addr:127.0.0.1  Mask:255.0.0.0
          UP LOOPBACK RUNNING  MTU:16436  Metric:1
          RX packets:0 errors:0 dropped:0 overruns:0 frame:0
          TX packets:0 errors:0 dropped:0 overruns:0 carrier:0
          collisions:0 txqueuelen:0
          RX bytes:0 (0.0 B)  TX bytes:0 (0.0 B)

wlan0     Link encap:Ethernet  HWaddr 00:0f:53:a0:04:57
          inet addr:192.168.1.13  Bcast:192.168.255.255  Mask:255.255.0.0
          UP BROADCAST RUNNING MULTICAST  MTU:1500  Metric:1
          RX packets:38 errors:0 dropped:0 overruns:0 frame:0
          TX packets:28 errors:0 dropped:0 overruns:0 carrier:0
          collisions:0 txqueuelen:1000
          RX bytes:6661 (6.5 KiB)  TX bytes:6377 (6.2 KiB)
```

Looking at the results of running this ifconfig, you can see that the Pi in question is connected by both a wired connection (eth0) with an IP address of 192.168.1.16, and a wireless one (wlan0) with an IP address of 192.168.1.13.

Discussion

As you can see, a Raspberry Pi can have more than one IP address (i.e., one for each network connection). So if you have both a wired connection and a wireless connection to your Pi, it would have two IP addresses. Normally, however, you would only connect it by one method or the other, not both.

Another way to find your IP address is to connect to the management console of your home hub and find the LAN page and then the IP table. There should be a device listed called *raspberrypi* with its IP address next to it.

See Also

Wikipedia (*http://bit.ly/13ZsgoY*) has everything you want to know about IP addresses.

2.3. Setting a Static IP Address

Problem

You would like to set the IP address of your Raspberry Pi so that it does not change.

Solution

To set the IP address of your Raspberry Pi, whether using a wired or wireless network, you need to edit the configuration file */etc/network/interfaces*.

If you view your */etc/network/interfaces* file using the following command:

```
$ more /etc/network/interfaces
```

it should look something like this:

```
iface lo inet loopback

iface eth0 inet dhcp

iface wlan0 inet dhcp
    wpa-ssid "ssidgoeshere"
    wpa-psk "passwordgoeshere"
```

To edit this file, type the following command:

```
$ sudo nano /etc/network/interfaces
```

If you are using a wired network, change the section for the adapter eth0; for a wireless network, change the section for wlan.

First, decide on an IP address to use. You need to pick one that is both unused by any other machine on the network and within the allowed range of IP addresses for your home hub.

Modify the contents of the file, changing the word dhcp to static and adding the following lines:

```
address 192.168.1.16
netmask 255.255.255.0
gateway 192.168.1.1
```

With the file changed as shown here, the static IP address of 192.168.1.16 has been assigned to the eth0 interface.

```
iface lo inet loopback

iface eth0 inet static
    address 192.168.1.16
    netmask 255.255.255.0
    gateway 192.168.1.1
```

```
iface wlan0 inet dhcp
    wpa-ssid "ssidgoeshere"
    wpa-psk "passwordgoeshere"
```

For most networks, the netmask setting should be set to 255.255.255.0 and the gateway should be the IP address of your home hub itself. This will be the same as the IP address you use to connect to its admin console.

After you edit and save the file, restart your Pi for the changes to take effect.

Discussion

Internal IP addresses are typically something like 192.168.1.16, where just the last number is changed for each of the different computers. Another common format for internal IP addresses is 10.0.0.16.

See Also

Wikipedia (*http://bit.ly/13ZsgoY*) has everything you want to know about IP addresses.

2.4. Setting the Network Name of a Raspberry Pi

Problem

You want to change the network name of your Raspberry Pi so that it's not just called "raspberrypi."

Solution

Changing the name of your Pi is pretty straightforward. There are just two files that need to be changed.

First, edit the file */etc/hostname*. You can do this by opening a Terminal window and typing the command:

```
$ sudo nano /etc/hostname
```

Replace "raspberrypi" with a name of your choice. This should remain one word, without any punctuation or unusual characters. This includes the _ character.

Second, open the file */etc/hosts* in an editor using the command:

```
$ sudo nano /etc/hosts
```

The file will look something like this:

```
127.0.0.1       localhost
::1             localhost ip6-localhost ip6-loopback
```

```
fe00::0      ip6-localnet
ff00::0      ip6-mcastprefix
ff02::1      ip6-allnodes
ff02::2      ip6-allrouters

127.0.1.1    raspberrypi
```

Change the text at the end that uses the old name ("raspberrypi") to the new name.

Restart the Pi, and you should find that the name has changed when you view it on the network from another computer.

Discussion

Changing the name of your Pi can be very useful, especially if you have more than one Pi connected to your network.

See Also

See also Recipe 2.3 to change the IP address of your Raspberry Pi.

2.5. Setting Up a Wireless Connection

Problem

You want to connect your Raspberry Pi to the Internet using a USB wireless adapter.

Solution

Setting up a wireless connection is very easy if you are using a recent version of the Raspbian distribution, as this includes a WiFi Config utility with a shortcut on the desktop. If you are not using a recent distribution, then you should update to one (see Recipes 1.4 through 1.8).

Simply plug a compatible USB wireless adapter (most are) into one of the USB sockets on your Raspberry Pi, launch the WiFi Config utility (Figure 2-3), and then click the Scan button to search for access points. Double-click on the access point (for your home hub) that you want to join and then enter the password in the PSK field.

Finally, click Connect to join the network.

Discussion

USB WiFi adapters can use quite a lot of power, so if you find your Pi rebooting unexpectedly or not booting properly, then you may need to use a larger power supply for it. Look for a supply that is 1.5A or more.

Figure 2-3. WiFi Config—scanning for networks

If you're using a keyboard and mouse as well, you may find you've run out of USB sockets. In that case, a USB hub is the answer. Choosing a hub with its own power supply will also help with the power problem.

See Also

For wired connections, see also Recipe 2.1. For a list of WiFi adapters that are compatible with the Raspberry Pi, go to *http://elinux.org/RPi_VerifiedPeripherals*.

2.6. Connecting with a Console Lead

Problem

No network connection is available, but you still want to be able to remotely control your Raspberry Pi from another computer.

Solution

Use a *console cable* to connect to a Raspberry Pi.

Console cables are great for a Pi that is going to be used *headless*—that is, without keyboard, mouse, or monitor. The console cable shown in Figure 2-4 is available from Adafruit (*http://www.adafruit.com/*).

Figure 2-4. A console cable

Connect the lead as follows:

1. Connect the red lead to the bottom-right pin, as shown in Figure 2-4.
2. Leave a gap of one space and connect the black lead to GND.
3. Connect the white lead to the left of the black lead.
4. Connect the green lead to the left of the white lead.

Note that the lead also provides 5V on the red lead, with enough power for the Pi on its own, but not with a lot of devices attached.

If you are a Windows or Mac user, you will need to install drivers for the USB lead, which are available for Windows (*http://bit.ly/1hAX6Mw*) and for Mac (*http://bit.ly/18oycat*).

To connect to the Pi from your Mac, you will need to run the Terminal and enter the command:

```
$ screen /dev/cu.PL2303-00001004 115200
```

The exact device name will be different, but if you press Tab after cu.P, it will auto-complete. After connecting, press Enter, and the Raspberry Pi login prompt should appear (Figure 2-5). The default username and password are *pi* and *raspberry*, respectively.

If you are trying to connect to your Raspberry Pi from a Windows computer, you need to download the terminal software called Putty (*http://www.putty.org*).

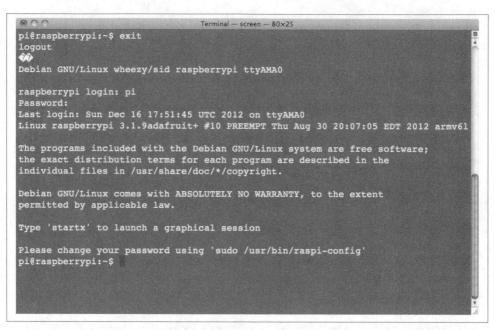

```
pi@raspberrypi:~$ exit
logout

Debian GNU/Linux wheezy/sid raspberrypi ttyAMA0

raspberrypi login: pi
Password:
Last login: Sun Dec 16 17:51:45 UTC 2012 on ttyAMA0
Linux raspberrypi 3.1.9adafruit+ #10 PREEMPT Thu Aug 30 20:07:05 EDT 2012 armv6l

The programs included with the Debian GNU/Linux system are free software;
the exact distribution terms for each program are described in the
individual files in /usr/share/doc/*/copyright.

Debian GNU/Linux comes with ABSOLUTELY NO WARRANTY, to the extent
permitted by applicable law.

Type 'startx' to launch a graphical session

Please change your password using 'sudo /usr/bin/raspi-config'
pi@raspberrypi:~$
```

Figure 2-5. Logging in with the console cable

When you run Putty, change the "Connection type" to Serial and set the speed to 115200.
You also need to set the "Serial line" to be the COM port in use by the cable. This may
be COM7, but if that does not work, check using the Windows Device manager.

When you click Open and press Enter, the Terminal session should start with a login
prompt.

Discussion

The console cable can be a very convenient way of using your Pi if you are travelling
light, as it provides both power and a way to control the Pi remotely.

See Also

You can find out more about using the serial console at this Adafruit tutorial (*http://
bit.ly/1bDL4vM*). Adafruit also sells console cables.

2.7. Controlling the Pi Remotely with SSH

Problem

You want to connect to a remote Pi from another computer using Secure Shell (SSH).

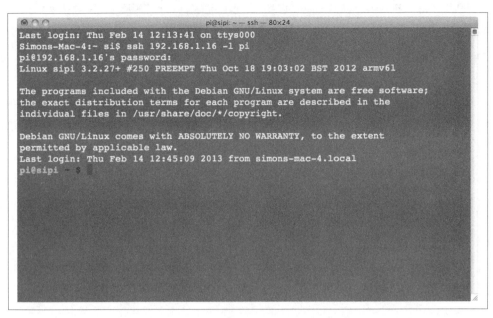

Figure 2-6. Logging in with SSH

Solution

The easiest way to set up SSH on the Raspberry Pi is to use the `raspi_config` application, which launches the first time that you boot up your Raspberry Pi with Raspbian. You can also start this at any time by entering the following command in Terminal:

```
$ sudo raspi-config
```

Scroll down to the SSH option and enable it.

> On newer versions of Raspbian, SSH is automatically enabled and there is no setting to change.

If you are using a Mac or have Linux installed on the computer from which you want to connect your Pi, then all you need to do to connect is open a Terminal window and enter the command:

```
$ ssh 192.168.1.12 -l pi
```

where the IP address is the IP address of your Pi (see Recipe 2.2). You will be prompted for your password and then logged in to the Pi (Figure 2-6).

To connect from a Windows computer, you will need to use Putty (Figure 2-5) to start an SSH session.

Discussion

SSH is a very common way of connecting to remote computers; any commands that you could issue on the Pi itself, you can use from the secure shell. It is also, as the name suggests, secure since the communication is encrypted.

Perhaps the only drawback is that it is a command-line rather than graphical environment. If you need access to the full Raspberry Pi desktop environment remotely, then you need to use VNC (Recipe 2.8).

See Also

See also this Adafruit tutorial (*http://bit.ly/HRnse4*).

2.8. Controlling the Pi Remotely with VNC

Problem

You want access to the full Raspbian graphical desktop of your Pi from a PC or Mac.

Solution

Install a VNC (Virtual Network Connection) server.

Open a Terminal session (or SSH session) on the Pi and run the following commands:

```
$ sudo apt-get update
$ sudo apt-get install tightvncserver
```

Having installed the VNC server, run it using the command:

```
$ vncserver :1
```

The first time you run this, you will be prompted to create a new password, so anyone connecting remotely has to enter the password before being granted access to the Pi.

To connect to the Pi from a remote computer, you will need to install a VNC client. RealVNC (*http://www.realvnc.com*) is a popular choice and connects well with *tightvnc*.

When you run the client program on your Mac or PC, you will be asked to enter the IP address of the VNC server you want to connect to (the IP address of your Pi). Enter ":1" after the IP address to indicate that you wish to connect to display number 1.

You will then be prompted for the password. Remember, this is the password you set previously after installing *tightvncserver* (Figure 2-7) and not necessarily the same as your normal Raspberry Pi password.

Figure 2-7. Logging in with VNC

Discussion

Although you can do most things with SSH, sometimes it is useful to have access to the graphical environment of the Pi.

If you want the VNC server to automatically start whenever you restart your Raspberry Pi, follow these steps:

```
$ cd /home/pi
$ cd .config
$ mkdir autostart
$ cd autostart
$ nano tightvnc.desktop
```

Paste the following into the editor window:

```
[Desktop Entry]
Type=Application
Name=TightVNC
Exec=vncserver :1
StartupNotify=false
```

As long as your Raspberry Pi is set to automatically log in and boot into the windowing environment, then VNC server will automatically start when you reboot.

See Also

See also this Adafruit tutorial (*http://bit.ly/172gOL7*).

2.9. File Sharing on a Mac Network

Problem

You want your Raspberry Pi to appear on the list of computers on your Mac's Finder so that you can connect to it and browse the filesystem using Finder.

Solution

The Mac OS X operating system includes built-in support for file viewing in Finder over the network (Figure 2-8). However, you must make some configuration changes on your Raspberry Pi for the Mac to pick these up.

Figure 2-8. Raspberry Pi on the Mac Finder

You will need to know the IP address of your Raspberry Pi (Recipe 2.2).

Now, on the Raspberry Pi, install `netatalk` using the command:

```
$ sudo apt-get install netatalk
```

Then, back on your Mac, in the Finder menu, select Go→Connect to Server and enter **afp://192.168.1.16** as the server address (but use the IP address for your Raspberry Pi instead of the one shown here). Then click Connect. You will be prompted to log in. I had to reboot my Raspberry Pi before I got the login prompt.

Log in using the name *pi* and your password, which will be *raspberry* by default. The Finder should then show you the contents of your home directory on the Raspberry Pi.

Now there are a few more configuration changes to make on the Raspberry Pi.

```
$ sudo apt-get install avahi-daemon
$ sudo update-rc.d avahi-daemon defaults
```

Next type the command:

```
$ sudo nano /etc/avahi/services/afpd.service
```

Paste the following code into the file.

```
<?xml version="1.0" standalone='no'?><!--*-nxml-*-->
<!DOCTYPE service-group SYSTEM "avahi-service.dtd">
<service-group>
   <name replace-wildcards="yes">%h</name>
   <service>
      <type>_afpovertcp._tcp</type>
      <port>548</port>
   </service>
</service-group>
```

To set the daemon running, type the command:

```
$ sudo /etc/init.d/avahi-daemon restart
```

Switch back to your Mac, and you should now see your Raspberry Pi in the Finder.

Discussion

Being able to move files easily between your Mac and your Raspberry Pi is very useful. It means that you can use files on your Pi without having to hook up a separate keyboard, mouse, and monitor.

You can also open files on the Raspberry Pi as if they were on your Mac. This has the advantage that you can edit them with TextMate or your favorite Mac text editor.

If you're using Windows or Linux, you can also share files by configuring your Raspberry Pi to work as NAS (Network Attached Storage); see Recipe 2.11.

See Also

The instructions here were adapted from this tutorial (*http://bit.ly/HRnOBv*), which credits Matt Richardson and Shawn Wallace's book *Getting Started with Raspberry Pi* (O'Reilly) as the original source.

2.10. Sharing the Pi Screen on a Mac

Problem

You have set up VNC, but would like to share the screen on your Raspberry Pi as if it were another Mac on your network.

Solution

First, follow Recipe 2.8 to install VNC. You will also need to complete Recipe 2.9.

Then enter the command:

```
$ sudo nano /etc/avahi/services/rfb.service
```

and paste the following into the editor:

```
<?xml version="1.0" standalone='no'?>
<!DOCTYPE service-group SYSTEM "avahi-service.dtd">
<service-group>
  <name replace-wildcards="yes">%h</name>
  <service>
    <type>_rfb._tcp</type>
    <port>5901</port>
  </service>
</service-group>
```

Now enter the command:

```
$ sudo /etc/init.d/avahi-daemon restart
```

You should now be able to see the Share Screen option shown in Figure 2-9. When prompted for a password, use the password you set up for VNC, not your general Raspberry Pi password.

Figure 2-9. Raspberry Pi screen sharing with the Mac Finder

Discussion

This recipe just adds a little convenience to the process of sharing the screen of your Raspberry Pi.

If you have more than one Raspberry Pi on your network, you need to give them different names so that you can identify them on the network (Recipe 2.4).

If you are a Windows or Linux user, you can still connect to a Raspberry Pi using VNC (Recipe 2.8).

See Also

The instructions here were adapted from this tutorial (*http://bit.ly/HRnOBv*), which credits Matt Richardson and Shawn Wallace's book *Getting Started with Raspberry Pi* (O'Reilly) as the original source.

2.11. Using a Raspberry Pi for Network Attached Storage

Problem

You want to use your Raspberry Pi as Network Attached Storage (NAS) by accessing a large USB drive attached to your Raspberry Pi from computers on your network.

Solution

The solution to this problem is to install and configure Samba. To do this, issue the commands:

```
$ sudo apt-get install samba
$ sudo apt-get install samba-common-bin
```

Now, attach the USB hard drive to the Raspberry Pi. It will automatically mount in your */media* folder. To check that it's there, use the command:

```
$ cd /media
$ ls
```

The drive should be listed with whatever name you gave it when you formatted it. It will automatically mount itself whenever the Raspberry Pi reboots.

You now need to configure Samba so the drive can be shared on the network. To do this, you first need to add a Samba user (*pi*). Enter the following command and type in a password:

```
$ sudo smbpasswd -a pi
New SMB password:
Retype new SMB password:
Added user pi.
```

You now need to make some changes to the file */etc/samba/smb.conf*, so enter the command:

```
$ sudo nano /etc/samba/smb.conf
```

The first line you're looking for is near the top of the file:

```
workgroup = WORKGROUP
```

You only need to change this if you plan to connect from a Windows machine. This should be the name of your Windows workgroup. For Windows XP, the default is MSHOME; for newer versions of Windows, it is HOME. (But check on your Windows network.)

The next change to be made is further down the file in the Authentication section. Find the line:

```
# security = user
```

Remove the # from the front to turn security on.

Finally, scroll right to the end of the file and add the following lines:

```
[USB]
path = /media/NAS
comment = NAS Drive
valid users = pi
writeable = yes
browseable = yes
create mask = 0777
public = yes
```

Save the file and then restart Samba by entering the command:

```
$ sudo /etc/init.d/samba restart
```

If all is well, your USB drive should now be shared on your network.

Discussion

To connect to the drive on a Mac, just select Go→Connect to Server from the Finder menu. Then enter **smb://raspberrypi/USB** in the Server Address field. You will see a login dialog box, where you need to change the username to *pi* (Figure 2-10).

Figure 2-10. Connecting to NAS with the Mac Finder

If you are connecting to the NAS from a Windows machine, then the exact procedure will vary depending on your version of Windows. However, the basic principle is that at some point, you will need to enter the network address, which should be *raspberrypi**USB* (Figure 2-11).

You will then be prompted for the username and password before you can use the NAS disk (Figure 2-12).

If you are a Linux user, this command should mount the NAS drive for you:

```
$ sudo mkdir /pishare
$ sudo smbmount -o username=pi,password=raspberry //192.168.1.16/USB /pishare
```

See Also

You may wish to change your Raspberry Pi's network name to something inappropriate like *piNAS* (see Recipe 2.4).

2.12. Network Printing

Problem

You want to print to a network printer using your Raspberry Pi.

Solution

Use CUPS (Common Unix Printing System).

Start by entering the following command into a Terminal to install CUPS. This may take some time.

```
$ sudo apt-get install cups
```

Figure 2-11. Connecting to NAS from Windows

Give yourself admin privileges for CUPS by entering the following command:

```
$ sudo usermod -a -G lpadmin pi
```

CUPS is configured via a web interface. Neither the Midori or Dillo web browsers work very well with the CUPS administration pages, but Iceweasel works just fine. You can install the Iceweasel browser by entering the following command:

```
$ sudo apt-get install iceweasel
```

Start Iceweasel from the Internet group of your Start menu and navigate to the address *http://localhost:631*.

Go to the Administration tab and choose the Add Printer option. This will display a list of printers that are on the network or connected directly to the Raspberry Pi's USB port (Figure 2-13).

Follow the series of dialogs to set up the printer.

Figure 2-12. Browsing NAS on Windows

Figure 2-13. Discovering printers with CUPS

Discussion

When you're finished, you can test out the printer by firing up AbiWord (see Recipe 4.2). Type some text and when you're ready to print it, you should see your newly added printer available for printing (Figure 2-14).

Figure 2-14. Printing from AbiWord

See Also

See also the official CUPS website (*http://www.cups.org*).

Operating System

3.0. Introduction

This chapter explores many aspects of the Linux operating system used by the Raspberry Pi. A lot of this involves the use of the command line.

3.1. Moving Files Around Graphically

Problem

You want to move files around using a graphical interface like you can on a Mac or PC.

Solution

Use the File Manager.

You can find this program on your Start menu in the Accessories group (Figure 3-1).

Figure 3-1. The File Manager

Discussion

The lefthand side of the File Manager shows the volumes that are mounted, so if you connect a USB flash drive or external USB drive, it will appear here.

The central area displays the files in the current folder, which you can navigate using the buttons in the toolbar, or by typing a location in the file path area at the top.

Right-click a file to reveal options that can be used on that file (Figure 3-2).

Figure 3-2. The File Manager

See Also

See also Recipe 3.4.

3.2. Starting a Terminal Session

Problem

When using a Raspberry Pi, you need to issue text commands in a Terminal.

Solution

Click on the LX Terminal icon on your desktop (Figure 3-3). If there is no shortcut on the desktop, you can also open a LX Terminal from the Start menu in the Accessories group.

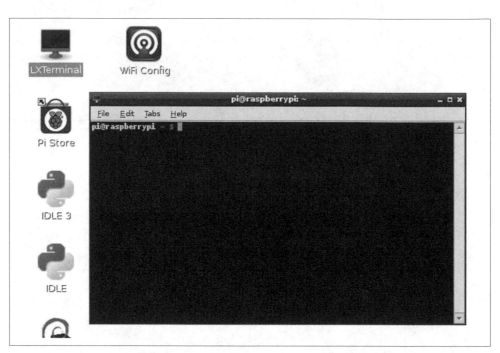

Figure 3-3. Opening LX Terminal

Discussion

When the LX Terminal starts, it will be set to your home directory (*/home/pi*).

You can open as many Terminal sessions as you want. It is often useful to have a couple open in different directories so that you don't have to constantly switch directories using cd (Recipe 3.3).

See Also

In the next section—Recipe 3.3—we will look at navigating the directory structure using the Terminal.

3.3. Navigating the Filesystem Using a Terminal

Problem

You need to know how to change directory and move about the fileystem using the Terminal.

Solution

The main command used for navigating the filesystem is cd (change directory). After cd you have to specify the directory that you want to change to. This can either be a *relative* path to a directory inside your current directory, or an *absolute* path to somewhere else on the filesystem. See "Discussion" on page 57.

To see what the current directory is, you can use the command pwd (print working directory).

Discussion

Try out a few examples. Open a Terminal session, and you should see a prompt like this:

```
pi@raspberrypi ~ $
```

The prompt that you will see after each command (pi@raspberrypi ~ $) is a reminder of your username (pi) and your computer name (raspberrypi). The ~ character is shorthand for your home directory (*/home/pi*). So, at any point, you can change your current directory to your home directory as follows:

```
$ cd ~
```

 Throughout the book, I use a $ at the start of each line where you are expected to type a command. The response from the command line will not be prefixed by anything; it will appear just as it does on the Raspberry Pi's screen.

You can confirm that the command did indeed set the directory to the home directory using the pwd command:

```
$ pwd
/home/pi
```

If you want to move up one level in the directory structure, you can use the special value .. (two dots) after the cd command, as shown here:

```
$ cd ..
$ pwd
/home
```

As you may have deduced by now, the path to a particular file or directory is made up of words separated by a /. So the very root of the whole filesystem is /, and to access the home directory within / you would refer to */home/*. Then, to find the *pi* directory within that, you would use */home/pi/*. The final / can be omitted from a path.

Paths can also be absolute (starting with a / and specifying the full path from the root), or they can be relative to the current working directory, in which case they must not start with a /.

You will have full read and write access to the files in your home directory, but once you move into the places where system files and applications are kept, your access to some files will be restricted to read only. You can override this (Recipe 3.11), but some care is required.

Check out the *root* of the directory structure by entering the following commands:

```
$ cd /
$ ls
bin   dev  home  lost+found  mnt  proc  run   selinux  sys  usr
boot  etc  lib   media       opt  root  sbin  srv      tmp  var
```

The `ls` command (list) shows us all the files and directories underneath / the root directory. You will see that there is a *home* directory listed, which is the directory you have just come from.

Now change into one of those directories by using the command:

```
$ cd etc
$ ls
adduser.conf            hosts.deny      polkit-1
alternatives            hp              profile
apm                     iceweasel       profile.d
apparmor.d              idmapd.conf     protocols
apt                     ifplugd         pulse
asound.conf             init            python
```

You will notice a couple of things. First, there are a lot of files and folders listed, more than can fit on the screen at once. You can use the scroll bar on the side of the Terminal window to move up and down.

Second, you will see that the files and folders have some color-coding. Files are displayed in white, while directories are blue.

Unless you particularly like typing, the Tab key offers a convenient short cut. If you start typing the name of a file, pressing the Tab key allows the autocomplete feature to attempt to complete the filename. For example, if you're going to change directory to *network*, type the command `cd netw` and then press the Tab key. Because *netw* is enough to uniquely identify the file or directory, pressing the Tab key will autocomplete it.

If what you have typed is not enough to uniquely identify the file or directory name, then pressing the Tab key another time will display a list of possible options that match what you have typed so far. So, if you had stopped at *net* and pressed the Tab key, you would see the following:

```
$ cd net
netatalk/ network/
```

You can provide an extra argument after `ls` to narrow down the things you want to list. Change directory to *etc* and then run the following:

```
$ ls f*
fake-hwclock.data   fb.modes   fstab   fuse.conf

fonts:
conf.avail   conf.d   fonts.conf   fonts.dtd

foomatic:
defaultspooler   direct   filter.conf

fstab.d:
pi@raspberrypi /etc $
```

The `*` character is called a wildcard. In specifying `f*` after `ls`, we are saying that we want to list everything that starts with an *f*.

Helpfully, the results first list all the files within *etc* that start with *f*, and then the contents of all the directories in that folder starting with *f*.

A common use of wildcards is to list all files with a certain extension (e.g., `ls *.docx`).

A convention in Linux (and many other operating systems) is to prefix files that should be hidden from the user by starting their name with a period. Any so-named files or folders will not appear when you type `ls` unless you also supply `ls` with the option `-a`. For example:

```
$ cd ~
$ ls -a
.                                   Desktop              .pulse
..                                  .dillo               .pulse-cookie
Adafruit-Raspberry-Pi-Python-Code   .dmrc                python_games
.advance                            .emulationstation    sales_log
.AppleDB                            .fltk                servo.py
.AppleDesktop                       .fontconfig          .stella
.AppleDouble                        .gstreamer-0.10      stepper.py.save
Asteroids.zip                       .gvfs                switches.txt.save
atari_roms                          indiecity            Temporary Items
.bash_history                       .local               thermometer.py
.bash_logout                        motor.py             .thumbnails
.bashrc                             .mozilla             .vnc
.cache                              mydocument.doc       .Xauthority
.config                             Network Trash Folder .xsession-errors
.dbus                               .profile             .xsession-errors.old
```

As you can see, most of the files and folders in your home directory are hidden.

See Also

See also Recipe 3.11.

3.4. Copying a File or Folder

Problem

You want to copy a file using a Terminal session.

Solution

Use the `cp` command to copy files and directories.

Discussion

You can, of course, copy files using the File Manager and its copy and paste menu options (Recipe 3.1).

The simplest example of copying in a Terminal session is to make a copy of a file within your working directory. The `cp` command is followed first by the file to copy, and then by the name to be given to the new file.

For example, the following example creates a file called *myfile.txt* and then makes a copy of it with the name *myfile2.txt*. You can find out more about the trick of creating a file using > in Recipe 3.8.

```
$ echo "hello" > myfile.txt
$ ls
myfile.txt
$ cp myfile.txt myfile2.txt
$ ls
myfile.txt     myfile2.txt
```

Although in this example, both file paths are local to the current working directory, the file paths can be to anywhere in the filesystem where you have write access. The following example copies the original file to an area */tmp*, which is a location for temporary files. Do not put anything important in that folder.

```
$ cp myfile.txt /tmp
```

Note that in this case, the name to be given to the new file is not specified, just the directory where it is to go. This will create a copy of *myfile.txt* in */tmp* with the same name of *myfile.tmp*.

Sometimes, rather than copying just one file, you may want to copy a whole directory full of files and possibly other directories. To copy such a directory, you need to use the `-r` option (for *recursive*). This will copy the directory and all its contents.

```
$ cp -r mydirectory mydirectory2
```

Whenever you are copying files or folders, if you do not have permission, the result of the command will tell you that. You will need to either change the permissions of the

folder into which you are copying (Recipe 3.12) or copy the files with superuser privileges (Recipe 3.11).

See Also

See also Recipes 3.5 and 3.12.

3.5. Renaming a File or Folder

Problem

You need to rename a file using a Terminal session.

Solution

Use the mv command to rename files and directories.

Discussion

The mv (*move*) command is used in a similar way to the cp command, except that the file or folder being moved is simply renamed rather than a duplicate being made.

For example, to simply rename a file from *my_file.txt* to *my_file.rtf* you would use the command:

```
$ mv my_file.txt my_file.rtf
```

Changing a directory name is just as straightforward, and you don't need the recursive -r option you used when copying in Recipe 3.4 since changing a directory's name implicitly means that everything within it is contained in a renamed directory.

See Also

See also Recipes 3.4 and 3.12.

3.6. Editing a File

Problem

You want to run an editor from the command line to change a config file.

Solution

Use the editor *nano* included with most Raspberry Pi distributions.

Discussion

To use nano, simply type the command nano followed by the name or path to the file that you want to edit. If the file does not exist, it will be created when you save it from the editor. However, this will only happen if you have write permissions in the directory where you are trying to write the file.

From your home directory, type the command **nano my_file.txt** to edit or create the file *nano my_file.txt*. Figure 3-4 shows *nano* in action.

Figure 3-4. Editing a file with nano

You cannot use the mouse to position the cursor; use the arrow keys instead.

The area at the bottom of the screen lists a number of commands that you access by holding down the Ctrl key and pressing the letter indicated. Most of these are not that useful. The ones that you are likely to use most of the time are:

Ctrl-X
> Exit. You will be prompted to save the file before nano exits.

Ctrl-V
> Next page. Think of it as an arrow pointing downward. This allows you to move through a large file a screen at a time.

Ctrl-Y
> Previous page.

Ctrl-W
> Where is. This allows you to search for a piece of text.

There are also some fairly crude cut-and-paste type options there, but in practice, it's easier to use the normal clipboard based on the menu that you access with a right-click (Figure 3-5).

Figure 3-5. Using the clipboard in nano

Using this clipboard also allows you to copy and paste text between other windows such as your browser.

When you're ready to save your changes to the file and exit nano, use the command Ctrl-X. Type **Y** to confirm that you want to save the file. nano then displays the filename as the default name to save the file under. Press Enter to save and exit.

If you want to abandon changes you have made, enter N in place of Y.

See Also

Editors are very much a matter of personal taste. Many other editors that are available for Linux will work just fine on Raspberry Pi. The *vim* (vi improved) editor has many

fans in the Linux world. This is also included in the popular Raspberry Pi distributions. It is not, however, an easy editor for the beginner. You can run it in the same way as nano, but using the command vi instead of nano. There are more details on using vim at *http://newbiedoc.sourceforge.net/text_editing/vim.html.en*.

3.7. Viewing the Contents of a File

Problem

You want to view the contents of a small file without editing it.

Solution

Use the cat or more commands to view the file.

For example:

```
$ more myfile.txt
This file contains
some text
```

Discussion

The cat command displays the whole contents of the file, even if it is longer than will fit on the screen.

The more command just displays one screen of text at a time. Press the space bar to display the next screen.

See Also

You can also use cat to concatenate (join together) a number of files (Recipe 3.28).

Another popular command related to more is less. less is like more except it allows you to move backward in the file as well as forward.

3.8. Creating a File Without Using an Editor

Problem

You want to create a one-line file without having to use an editor.

Solution

Use the > and echo commands to redirect what you type on the command line to a file.

For example:

```
$ echo "file contents here" > test.txt
$ more test.txt
file contents here
```

 The > command overwrites any existing file, so use it with caution.

Discussion

This can be useful for quickly creating a file.

See Also

To use the `cat` command to view files without using an editor, see Recipe 3.7.

To use > to capture other kinds of system output, see Recipe 3.27.

3.9. Creating a Directory

Problem

You want to create a new directory using the Terminal.

Solution

The `mkdir` command will create a new directory.

Discussion

To create a directory, use the `mkdir` command. Try out the following example:

```
$ cd ~
$ mkdir my_directory
$ cd my_directory
$ ls
```

You need to have write permission in the directory within which you are trying to create the new directory.

See Also

For general information on using the Terminal to navigate the filesystem, see Recipe 3.3.

3.10. Deleting a File or Directory

Problem

You want to delete a file or directory using the Terminal.

Solution

The `rm` (remove) command will delete a file or directory and its contents. It should be used with extreme caution.

Discussion

Deleting a single file is simple and safe. The following example will delete the file *my_file.txt* from the home directory:

```
$ cd ~
$ rm my_file.txt
$ ls
```

You need to have write permission in the directory within which you are trying to carry out the deletion.

You can also use the * wildcard when deleting files. This example will delete all the files starting with *my_file.* in the current directory:

```
$ rm my_file.*
```

You could also delete all the files in the directory by typing:

```
$ rm *
```

If you want to recursively delete a directory and all its contents, including any directories that it contains, you can use the `-r` option:

```
$ rm -r mydir
```

 When deleting files from a Terminal window, remember that you do not have the safety net of a recycle bin from which files can be undeleted. Also, generally speaking, you won't be given the option to confirm, the files will just be immediately deleted. This can be totally devastating if you combine it with the command `sudo` (Recipe 3.11).

See Also

See also Recipe 3.3

If you are concerned about accidentally deleting files or folders, you can force the `rm` command to confirm by setting up a command alias (Recipe 3.32).

3.11. Performing Tasks with Superuser Privileges

Problem

Some commands don't work because you have *insufficient privileges*. You need to issue commands with superuser privileges.

Solution

The sudo (superuser do) command allows you to perform actions with superuser privileges. Just prefix the command with sudo.

Discussion

Most tasks that you want to perform on the command line can usually be performed without superuser privileges. The most common exceptions to this are when you're installing new software and editing configuration files.

For example, if you try to use the command apt-get update, you will receive a number of permission denied messages:

```
$ apt-get update
E: Could not open lock file /var/lib/apt/lists/lock - open (13: Permission denied)
E: Unable to lock directory /var/lib/apt/lists/
E: Could not open lock file /var/lib/dpkg/lock - open (13: Permission denied)
E: Unable to lock the administration directory (/var/lib/dpkg/), are you root?
```

The message at the end--are you root?--gives the game away. If you issue the same command prefixed with sudo, the command will work just fine:

```
$ sudo apt-get update
Get:1 http://mirrordirector.raspbian.org wheezy InRelease [12.5 kB]
Hit http://archive.raspberrypi.org wheezy InRelease
Get:2 http://mirrordirector.raspbian.org wheezy/main Sources [6,241 kB]
Hit http://archive.raspberrypi.org wheezy/main armhf Packages
Ign http://archive.raspberrypi.org wheezy/main Translation-en_GB
Ign http://archive.raspberrypi.org wheezy/main Translation-en
40% [2 Sources 2,504 kB/6,241 kB 40%]
```

See Also

To understand more about file permissions, see Recipe 3.12.

To install software using apt-get, see Recipe 3.16.

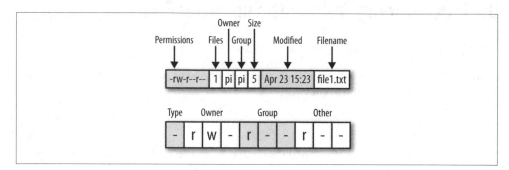

Figure 3-6. File permissions

3.12. Understanding File Permissions

Problem

You have seen the strange characters that accompany a filename when it is listed. You would like to know what they all mean.

Solution

To see the permissions and ownership information relating to files and directories, use the ls command with the option -l.

Discussion

Run the command ls -l, and you will see a result like this:

```
$ ls -l
total 16
-rw-r--r-- 1 pi pi    5 Apr 23 15:23 file1.txt
-rw-r--r-- 1 pi pi    5 Apr 23 15:23 file2.txt
-rw-r--r-- 1 pi pi    5 Apr 23 15:23 file3.txt
drwxr-xr-x 2 pi pi 4096 Apr 23 15:23 mydir
```

Figure 3-6 shows the different sections of the listing information. The first section contains the permissions. In the second column, the number 1 (labeled "Files") indicates how many files are involved. This field only makes sense if the listing entry is for a directory; if it is a file, it will mostly be just 1. The next two entries (both pi) are the owner and group of the file. The size entry (the fifth column) indicates the size of the file in bytes. The modified date will change every time the file is edited or changed and the final entry is the actual name of the file or directory.

The permissions string is split into four sections (Type, Owner, Group, and Other). The first section is the type of the file. If this is a directory, it will be the character *d*; if it is just a file, the entry will be just a -.

The next section of three characters specifies the owner permissions for that file. Each character is a flag that is either on or off. So if the owner has read permissions, there will be a *r* in the first character position. If he has write permissions, there will be a *w* in the second slot. The third position, which is - in this example, can be *x* if the file is executable (a program or script) for the owner.

The third section has the same three flags but for any users in the group. Users can be organized into groups. So, in this case, the file has a user *pi* and a group ownership of *pi*. So, if there were any other users in the group *pi*, they would have the permissions specified here.

The final section specifies the permissions for any other users who are neither *pi* nor in the group *pi*.

Since most people will only ever use the Raspberry Pi as the user *pi*, the permissions of most interest are in the first section.

See Also

To change file permissions, see Recipe 3.13.

3.13. Changing File Permissions

Problem

You need to change the permissions on a file.

Solution

The command chmod is used to modify file permissions.

Discussion

Common reasons why you might want to change file permissions include needing to edit a file that is marked as read-only and giving a file *execute* permissions so that it can run as a program or script.

The chmod command allows you to add or remove permissions for a file. There are two syntaxes for doing this; one requires the use of octal (base 8) and the other is text-based. We will use the easier-to-understand text method.

The first parameter to chmod is the change to make, and the second is the file or folder to which it should apply. This change parameter takes the form of the permission scope (+, -, = for add, remove, or set, respectively) and then the permission type.

For example, the following code will add execute (*x*) rights to the file for the owner of the file *file2.txt*.

```
$ chmod u+x file2.txt
```

If we now list the directory, we can see that the *x* permission has been added.

```
$ ls -l
total 16
-rw-r--r-- 1 pi pi    5 Apr 23 15:23 file1.txt
-rwxr--r-- 1 pi pi    5 Apr 24 08:08 file2.txt
-rw-r--r-- 1 pi pi    5 Apr 23 15:23 file3.txt
drwxr-xr-x 2 pi pi 4096 Apr 23 15:23 mydir
```

If we wanted to add execute permission for the group or other users, then we could use *g* and *o*, respectively. The letter *a* will add the permission to everyone.

See Also

For background on file permissions, see Recipe 3.12.

See Recipe 3.14 for changing file ownership.

3.14. Changing File Ownership

Problem

You need to change the ownership of a file.

Solution

The command chown (change owner) is used to modify the ownership of a file or directory.

Discussion

As we discovered in Recipe 3.12, any file or directory has both an owner and a group associated with it. Since most users of the Raspberry Pi will just have the single user of *pi*, we don't really need to worry about groups.

Occasionally, you will find files on your system that have been installed with a different user than *pi*. If this is the case, you can change the ownership of the file using the chown command.

To just change the owner of a file, use chown followed by the new owner and group, separated by a colon, and then the name of the file.

You will probably find that you need superuser privileges to change ownership, in which case, prefix the command with sudo (Recipe 3.11).

```
$ sudo chown root:root file2.txt
$ ls -l
total 16
-rw-r--r-- 1 pi    pi      5 Apr 23 15:23 file1.txt
-rwxr--r-- 1 root  root    5 Apr 24 08:08 file2.txt
-rw-r--r-- 1 pi    pi      5 Apr 23 15:23 file3.txt
drwxr-xr-x 2 pi    pi   4096 Apr 23 15:23 mydir
```

See Also

For background on file permissions, see Recipe 3.12.

Also see Recipe 3.13 for changing file permissions.

3.15. Making a Screen Capture

Problem

You want to capture an image of the Raspberry Pi's screen and save it to a file.

Solution

Install and use the delightfully named scrot screen capture software.

Discussion

To install scrot, run the following command from Terminal:

```
$ sudo apt-get install scrot
```

The simplest way to trigger a screen capture is to just enter the command scrot. This will immediately take an image of the primary display and save it in a file named something like *2013-04-25-080116_1024x768_scrot.png* within the current directory.

Sometimes you want a screenshot to show a menu being opened or something that generally disappears when the window in which you are interested loses focus. For such situations, you can specify a delay before the capture takes place using the -d option.

```
$ scrot -d 5
```

The delay is specified in seconds.

If you capture the whole screen, you can crop it later with image editing software, such as Gimp (Recipe 4.10). However, it is more convenient to just capture an area of the screen in the first place, which you can do using the -s option.

To use this option, type this command and then drag out the area of screen that you want to capture with the mouse.

```
$ scrot -s
```

The filename will include the dimensions in pixels of the image captured.

See Also

The scrot command has a number of other options to control things like using multiple screens and changing the format of the saved file. You can find out more about scrot from its manpage by entering the following command.

```
$ man scrot
```

For more information on installing with apt-get, see Recipe 3.16.

3.16. Installing Software with apt-get

Problem

You want to install software using the command line.

Solution

The most used tool for installing software from a Terminal session is apt-get.

The basic format of the command, which you must run as a superuser, is:

```
$ sudo apt-get install <name of software>
```

So, for example, to install the AbiWord word processing software, you would enter the command:

```
$ sudo apt-get install abiword
```

Discussion

The apt-get package manager uses a list of available software. This list is included with the Raspberry Pi operating system distribution that you use, but is likely to be out-of-date. So, if the software that you try to install is reported by apt-get as *not found*, run the following command to update the list:

```
$ sudo apt-get update
```

The list and the software packages for installation are all on the Internet, so none of this will work unless your Raspberry Pi has an Internet connection.

 If you find that when you update, you get an error like E: Problem with MergeList /var/lib/dpkg/status, try running these commands:

```
sudo rm /var/lib/dpkg/status
sudo touch /var/lib/dpkg/status
```

The installation process can often take a while because the files have to be downloaded and installed. Some installations will also add shortcuts to your desktop, or the program groups on your Start menu.

You can search for software to install using the command apt-get search followed by a search string such as abiword. This will then display a list of matching packages that you could install.

See Also

See Recipe 3.17 for removing programs that you no longer need so that you can free up space.

See also Recipe 3.19 for downloading source code from GitHub.

3.17. Removing Software Installed with apt-get

Problem

Having installed a whole load of programs using apt-get, you now find that you want to remove some of them.

Solution

The apt-get utility has an option (remove) that will remove a package, but only packages that have been installed with apt-get install.

For example, if you wanted to remove AbiWord, you would use the command:

```
$ sudo apt-get remove abiword
```

Discussion

Removing a package like this does not always delete everything, as packages often also have prerequisite packages that are installed as well. To remove these, you can use the autoremove option, as shown here:

```
$ sudo apt-get autoremove abiword
$ sudo apt-get clean
```

The `apt-get clean` option will do some further tidying up of unused package installation files.

See Also

See also Recipe 3.16 for installing packages using `apt-get`.

3.18. Fetching Files from the Command Line

Problem

You want to download a file from the Internet without using a web browser.

Solution

Use the `wget` command to fetch a file from the Internet.

For example:

```
$ wget http://www.icrobotics.co.uk/wiki/images/c/c3/Pifm.tar.gz
--2013-06-07 07:35:01--  http://www.icrobotics.co.uk/wiki/images/c/c3/Pifm.tar.gz
Resolving www.icrobotics.co.uk (www.icrobotics.co.uk)... 155.198.3.147
Connecting to www.icrobotics.co.uk (www.icrobotics.co.uk)|155.198.3.147|
:80... connected.
HTTP request sent, awaiting response... 200 OK
Length: 5521400 (5.3M) [application/x-gzip]
Saving to: `Pifm.tar.gz'

100%[=====================================================>] 5,521,400     601K/s

2013-06-07 07:35:11 (601 KB/s) - `Pifm.tar.gz' saved [5521400/5521400]
```

If your URL contains any special characters, it is a good idea to enclose it in double quotes. This example URL is from Recipe 4.9.

Discussion

You will find instructions for installing software that rely on using `wget` to fetch files. It is often more convenient to do this from the command line rather than use a browser, find the file, download it, and then copy it to the place you need it.

The `wget` command takes the URL to download as its argument and downloads it into the current directory. It's typically used to download an archive file of some type, but will also download any web page.

See Also

For more information on installing with `apt-get` see Recipe 3.16.

For information on selecting and using a browser, see Recipe 4.3.

3.19. Fetching Source Code with git

Problem

Sometimes Python libraries and other software are supplied via a Git repository URL. You need to be able to fetch them onto your Raspberry Pi.

Solution

To use code in Git repositories, you need to download Git and then use the `git clone` command to fetch the files.

Discussion

Git is a source code management system. To install it, use the command:

```
$ sudo apt-get install git
```

Once Git is installed, you can use the `clone` command to fetch the source code files you want. If you are doing this into a directory where you have write access, you don't need to do this as a superuser.

For example, the following command will fetch all the source code examples from this book:

```
$ git clone https://github.com/simonmonk/raspberrypi_cookbook.git
```

See Also

Find out more about Git (*http://www.git-scm.com*) and the Git hosting service GitHub (*http://www.github.com*).

See also Recipe 3.16.

3.20. Running a Program or Script Automatically on Startup

Problem

You want to arrange for a script or program to start automatically every time the Raspberry Pi reboots.

Solution

Debian Linux, upon which most Raspberry Pi distributions are based, uses a dependency-based mechanism for automating the running of commands at startup. This is a little tricky to use and involves creating a configuration file for the script or program that you want to run in a folder called *init.d*.

Discussion

The following example shows you how to run a Python script in your home directory. The script could do anything, but in this case, the script runs a simple Python web server, which is described further in Recipe 7.16.

The steps involved in this are:

1. Create an *init* script.
2. Make the *init* script executable.
3. Tell the system about the new *init* script.

First, create the *init* script. This needs to be created in the folder */etc/init.d/*. The script can be called anything, but in this example, we will call it my_server.

Create the new file using nano with the following command:

```
$ sudo nano /etc/init.d/my_server
```

Paste the following code into the editor window and save the file:

```
### BEGIN INIT INFO
# Provides: my_server
# Required-Start: $remote_fs $syslog $network
# Required-Stop: $remote_fs $syslog $network
# Default-Start: 2 3 4 5
# Default-Stop: 0 1 6
# Short-Description: Simple Web Server
# Description: Simple Web Server
### END INIT INFO

#! /bin/sh
```

```
# /etc/init.d/my_server

export HOME
case "$1" in
  start)
    echo "Starting My Server"
    sudo /usr/bin/python /home/pi/myserver.py  2>&1 &
  ;;
stop)
  echo "Stopping My Server"
  PID=`ps auxwww | grep myserver.py | head -1 | awk '{print $2}'`
  kill -9 $PID
  ;;
*)
  echo "Usage: /etc/init.d/my_server {start|stop}"
  exit 1
;;
esac
exit 0
```

This is quite a lot of work to automate the running of a script, but most of it is boiler-plate code. To run a different script, just work your way through the script, changing the descriptions and the name of the Python file you want to run.

The next step is to make this file executable for the owner, which you do using this command:

```
$ sudo chmod +x /etc/init.d/my_server
```

Now that the program is set up as a service, you can test that everything is OK, before you set it to autostart as part of the boot sequence, using the following command:

```
$ /etc/init.d/my_server start
Starting My Server
Bottle v0.11.4 server starting up (using WSGIRefServer())...
Listening on http://192.168.1.16:80/
Hit Ctrl-C to quit.
```

Finally, if that runs OK, use the following command to make the system aware of the new service that you have defined:

```
$ sudo update-rc.d my_server defaults
```

See Also

For more information on changing file and folder permissions, see Recipe 3.13.

3.21. Running a Program or Script Automatically at Regular Intervals

Problem

You want to run a script once a day or at regular intervals.

Solution

Use the Linux crontab command.

To do this, the Raspberry Pi needs to know the time and date, and therefore needs a network connection or a real-time clock. See Recipe 11.13.

Discussion

The command crontab allows you to schedule events to take place at regular intervals. This can be daily or hourly, and you can even define complicated patterns so different things happen on different days of the week. This is useful for backup tasks that you might want to run in the middle of the night.

You can edit the scheduled events using the following command:

```
$ crontab -e
```

If the script or program that you want to run needs to be run by a superuser, then prefix all the crontab commands with sudo (Recipe 3.11).

The comment line indicates the format of a crontab line. The digits are, in order, minute, hour, day of month, month, day of week, and then the command that you want to run.

If there is a * in the digit position, that means *every*; if there is a number there instead, the script will only be run at that minute/hour/day of the month.

So, to run the script every day at 1 a.m., you would add the line shown in Figure 3-7.

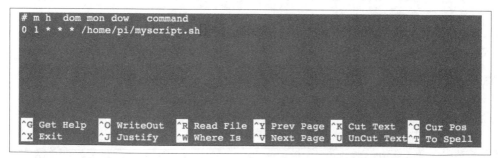

Figure 3-7. Editing crontab

Using only a single digit, you can specify ranges to run the script just on weekdays; for example:

```
0 1 * * 1-5 /home/pi/myscript.sh
```

If your script needs to be run from a particular directory, you can use a semicolon (;) to separate multiple commands as shown here:

```
0 1 * * * cd /home/pi; python mypythoncode.py
```

See Also

You can see the full documentation for crontab by entering this command:

```
$ man crontab
```

3.22. Finding Things

Problem

You want to find a file that you know is on the system somewhere.

Solution

Use the Linux find command.

Discussion

Starting with a directory specified in the command, the find command will search, and if it finds the file, display its location. For example:

```
$ find /home/pi -name gemgem.py
/home/pi/python_games/gemgem.py
```

You can start the search higher up the tree, even at the root of the whole filesystem (/). This will make the search take a lot longer, and will also produce error messages. You can redirect these error messages by adding **2>/dev/null** to the end of the line.

To search for the file throughout the entire filesystem, use the following command:

```
$ find / -name gemgem.py 2>/dev/null
/home/pi/python_games/gemgem.py
```

You can also use wildcards with find as follows:

```
$ find /home/pi -name match*
/home/pi/python_games/match4.wav
/home/pi/python_games/match2.wav
/home/pi/python_games/match1.wav
/home/pi/python_games/match3.wav
```

```
/home/pi/python_games/match0.wav
/home/pi/python_games/match5.wav
```

See Also

The find command has a number of other advanced features for searching. To see the full manpage documentation for find, use this command:

```
$ man find
```

3.23. Using the Command-Line History

Problem

You want to be able to repeat commands on the command line without having to type them again.

Solution

Use the up and down keys to select previous commands from the command history, and the history command with grep to find older commands.

Discussion

You can access the previous command you ran by pressing the up key. Pressing it again will take you to the command before that, and so on. If you overshoot the command you wanted, the down arrow key will take you back in the other direction.

If you want to cancel without running the selected command, use Ctrl-C.

Over time, your command history will grow too large to find a command that you used ages ago. To find a command from way back, you can use the history command.

```
$ history
    1  sudo nano /etc/init.d/my_server
    2  sudo chmod +x /etc/init.d/my_server
    3  /etc/init.d/my_server start
    4  cp /media/4954-5EF7/sales_log/server.py myserver.py
    5  /etc/init.d/my_server start
    6  sudo apt-get update
    7  sudo apt-get install bottle
    8  sudo apt-get install python-bottle
```

This lists all your command history and is likely to have far too many entries for you to find the one you want. To remedy this, you can *pipe* the history command into the grep command, which will just display results matching a search string. So, for example, to find all the apt-get (Recipe 3.16) commands that you've issued, you can use the line:

```
$ history | grep apt-get
    6  sudo apt-get update
    7  sudo apt-get install bottle
    8  sudo apt-get install python-bottle
   55  history | grep apt-get
```

Each history item has a number next to it, so if you find the line you were looking for, you can run it using ! followed by the history number, as shown here:

```
$ !6
sudo apt-get update
Hit http://mirrordirector.raspbian.org wheezy InRelease
Hit http://mirrordirector.raspbian.org wheezy/main armhf Packages
Hit http://mirrordirector.raspbian.org wheezy/contrib armhf Packages
.....
```

See Also

To find files rather than commands, see Recipe 3.22.

3.24. Monitoring Processor Activity

Problem

The Raspberry Pi can run a bit slow sometimes, so you want to see what is hogging the processor.

Solution

Use the Task Manager utility, which you'll find on the Start menu in the System Tools program group (Figure 3-8).

The Task Manager allows you to see at a glance how much CPU and memory are being used. You can also right-click on a process and select the option to kill it from the pop-up menu.

The graphs at the top of the window display the total CPU usage and memory. The processes are listed below that, and you can see what share of the CPU each is taking.

Discussion

If you prefer to do this type of thing from the command line, use the Linux top command to display very similar data about processor and memory and which processes are using the most resources (Figure 3-9). You can then use the kill command to kill a process. You will need to do this as superuser.

Figure 3-8. The Task Manager

Figure 3-9. Using the top command to see resource usage

In this case, you can see that the top process is a Python program that uses 97% of CPU. The first column shows its process ID (2447). To kill this process, enter this command:

```
$ kill 2447
```

It is quite possible to kill some vital operating system process this way, but if you do, powering off your Pi and turning it back on again will restore things to normal.

Sometimes, you may have a process running that is not immediately visible when you use top. If this is the case, you can search all the processes running by using the ps command and piping (Recipe 3.29) the results to the grep command, which will search the results and highlight items of interest.

For example, to find the process ID for our CPU-hogging Python process, we could run the following command:

```
$ ps -ef | grep "python"
pi        2447  2397 99 07:01 pts/0    00:00:02 python speed.py
pi        2456  2397  0 07:01 pts/0    00:00:00 grep --color=auto python
```

In this case, the process ID for the Python program *speed.py* is 2447. The second entry in the list is the process for the ps command itself.

A variation on the kill command is the killall command. Use this with caution, as it kills all processes that match its argument. So, for example, the following command will kill all Python programs running on the Raspberry Pi.

```
$ sudo killall python
```

See Also

See also the manpages for top, ps, grep, kill, and killall.

You can view these by typing *man* followed by the name of the command, as shown here:

```
$ man top
```

3.25. Working with File Archives

Problem

You have downloaded a compressed file and want to uncompress it.

Solution

Depending on the file type, you will need to use the tar or gunzip commands.

Discussion

If the file that you want to uncompress just has the extension *.gz*, you can unzip it using the command:

```
$ gunzip myfile.gz
```

You also often find files (called *tarballs*) that contain a directory that has been archived with the Linux tar utility and then compressed with gzip into a file with a name like *myfile.tar.gz*.

You can extract the original files and folders out of a tarball by using the tar command.

```
$ tar -xzf myfile.tar.gz
```

See Also

You can find out more about tar from the manpages, which you can access by using the command man tar.

3.26. Listing Connected USB Devices

Problem

You have plugged in a USB device and want to make sure Linux recognizes it.

Solution

Use the lsusb command. This will list all the devices attached to the USB ports on your Raspberry Pi:

```
$ lsusb
Bus 001 Device 002: ID 0424:9512 Standard Microsystems Corp.
Bus 001 Device 001: ID 1d6b:0002 Linux Foundation 2.0 root hub
Bus 001 Device 003: ID 0424:ec00 Standard Microsystems Corp.
Bus 001 Device 004: ID 15d9:0a41 Trust International B.V. MI-2540D [Optical mouse]
```

Discussion

This command will tell you whether a device is connected or not, but it will not guarantee that it is working correctly. There may be drivers to install or configuration changes to make for the hardware.

See Also

For an example of using lsusb when attaching an external webcam, see Recipe 4.5.

3.27. Redirecting Output from the Command Line to a File

Problem

You want to quickly create a file with some text, or record a directory listing into a file.

Solution

Use the > command to redirect output that would otherwise appear on the command line.

For example, to copy a directory listing into a file called *myfiles.txt*, do the following:

```
$ ls > myfiles.txt
$ more myfiles.txt
Desktop
indiecity
master.zip
mcpi
```

Discussion

You can use the > command on any Linux command that produces output, even if you are running, say, a Python program.

You can also use the opposite < command to redirect user input, although this is not nearly as useful as >.

See Also

To use > to join together a number of files, see Recipe 3.28.

3.28. Concatenating Files

Problem

You have a number of text files and you want to join them into one big file.

Solution

Use the cat command to *concatenate* a number of files into one output file.

For example:

```
$ cat file1.txt file2.txt file3.txt > full_file.txt
```

Discussion

Joining files is the real purpose of the cat command. You can supply as many filenames as you like, and they will all be written to the file you direct to. If you do not redirect the output, then it will just appear in your Terminal window. If they are big files, this might take some time!

See Also

See also Recipe 3.7, where `cat` is used to display the contents of a file.

3.29. Using Pipes

Problem

You want to use the output of one Linux command as the input to another command.

Solution

Use the `pipe` command, which is the *bar* symbol (|) on your keyboard, to `pipe` the output of one command to another. For example:

```
$ ls -l *.py | grep Jun
-rw-r--r-- 1 pi pi 226 Jun  7 06:49 speed.py
```

This example will find all the files with the extension *py* that also have *Jun* in their directory listing, indicating that they were last modified in June.

Discussion

At first sight, this looks rather like output redirection using > (Recipe 3.27). The difference is that +>_ will not work where the target is another program. It will only work for redirecting to a file.

You can chain together as many programs as you like, as shown here, although this isn't something you will do often:

```
$ command1 | command2 | command3
```

See Also

See also Recipe 3.24 for an example of using `grep` to find a process, and Recipe 3.23 to search your command history using a pipe and `grep`.

3.30. Hiding Output to the Terminal

Problem

You want to run a command, but you don't want the output filling up your screen.

Solution

Redirect the output to */dev/null* using >.

For example:

```
$ ls > /dev/null
```

Discussion

This example illustrates the syntax, but is otherwise pretty useless. A more common use is where you're running a program and the developer has left a lot of trace messages in her code, which you don't really want to see. The following example hides superfluous output from the find command (see Recipe 3.22).

```
$ find / -name gemgem.py 2>/dev/null
/home/pi/python_games/gemgem.py
```

See Also

For more information about redirecting standard output, see Recipe 3.27.

3.31. Running Programs in the Background

Problem

You want to run a program but also work on some other task.

Solution

Run the program or command in the background using the & command.

For example:

```
$ python speed.py &
[1] 2528
$ ls
```

Rather than wait until the program has finished running, the command line displays the process ID (the second number) and immediately allows you to continue with whatever other commands you want to run. You can then use this process ID to kill the background process (Recipe 3.24).

To bring the background process back to the foreground, use the fg command:

```
$ fg
python speed.py
```

This will report the command or program that is running and then wait for it to finish.

Discussion

Output from the background process will still appear in the Terminal.

An alternative to putting processes in the background is to just open more than one Terminal window.

See Also

For information on managing processes, see Recipe 3.24.

3.32. Creating Command Aliases

Problem

You want to create aliases to commands that you use frequently.

Solution

Edit the file ~/.bashrc using nano (Recipe 3.6), and then move to the end of the file and add as many lines as you want, like this:

```
alias l='ls -a'
```

This creates an alias called *l* that, when entered, will be interpreted as the command ls -a.

Save and exit the file using Ctrl-X and Y, and then to update the Terminal with the new alias, type the following command:

```
$ source .bashrc
```

Discussion

Many Linux users set up an alias for rm like the following, so that it confirms deletions.

```
$ alias rm='rm -i'
```

This is not a bad idea, as long as you do not forget when you use someone else's system that doesn't have this alias set up!

See Also

For more information about rm, see Recipe 3.10.

3.33. Setting the Date and Time

Problem

You want to manually set the date and time on your Raspberry Pi.

Solution

Use the Linux `date` command.

The date and time format is MMDDHHMMYYYY, where MM is the month number, DD the day of the month, HH and MM the hours and minutes, respectively, and YYYY the year.

For example:

```
$ sudo date 010203042013
Wed Jan  2 03:04:00 UTC 2013
```

Discussion

If the Raspberry Pi is connected to the Internet, then as it boots up it will automatically set its own time using an Internet time server.

You can also use `date` to display the UTC time by just entering `date` on its own:

```
$ date
Wed Jan  2 03:08:14 UTC 2013
```

See Also

If you want your Raspberry Pi to maintain the correct time, even when there is no network, then you can use an RTC (real-time clock) module (Recipe 11.13).

3.34. Finding Out How Much Room You Have on the SD Card

Problem

You want to know how much free space there is on the SD card.

Solution

Use the Linux `df` command:

```
$ df -h
Filesystem      Size  Used Avail Use% Mounted on
rootfs          3.6G  1.7G  1.9G  48% /
/dev/root       3.6G  1.7G  1.9G  48% /
devtmpfs        180M     0  180M   0% /dev
tmpfs            38M  236K   38M   1% /run
tmpfs           5.0M     0  5.0M   0% /run/lock
tmpfs            75M     0   75M   0% /run/shm
/dev/mmcblk0p1   56M   19M   38M  34% /boot
```

Discussion

Looking at the first line of the results, you can see that there is 3.6 GB of storage on the SD card, of which 1.7 GB is used.

When you run out of disk space, you are likely to get unexpected bad behavior, such as error messages saying that a file could not be written.

See Also

You can find the manpage for df using man df.

Software

4.0. Introduction

This chapter contains a number of recipes for using ready-made software on the Raspberry Pi.

Some of the recipes in this chapter are concerned with converting the Raspberry Pi into a single-use appliance, while the others use specific pieces of software on a Raspberry Pi.

4.1. Making a Media Center

Problem

You want to convert your Raspberry Pi into a super-duper media center.

Solution

A Raspberry Pi actually makes a pretty good media center. Figure 4-1 shows one running XBMC (Xbox Music Center).

Figure 4-1. Raspberry Pi as a media center

Raspberry Pi is perfectly capable of playing full HD video as well as streamed music, MP3 files, and Internet radio.

XBMC is an open source project that was originally developed to convert Xbox game consoles into media centers. The code has since been ported to many platforms including the Raspberry Pi.

Converting your Raspberry Pi to a media center is a matter of creating a new SD card that has a distribution including XBMC. We will use Raspbmc.

1. Download the disk image onto your computer.

 To create an SD card with the Raspbmc distribution on it, you need access to a PC (Windows, Mac, or Linux) with an SD card writer.

 Download the image file for Raspbmc from *http://www.raspbmc.com/download*.

 Scroll down the web page until you get to the section "Just want an image without a fancy installer?" and download the file labeled Standalone Image.

2. Copy the *image* file onto an SD card.

 To create an SD card with the Raspbmc image file, follow exactly the process described in Recipe 1.6. You should use an SD card of at least 4 GB.

3. Put the SD card into the Raspberry Pi and start it up.

 You may need to answer a few configuration questions during the first boot, and then you are on your way.

Discussion

XBMC is a powerful piece of software with many features. Perhaps the simplest way to check that it is working is to put some music and/or video files onto a USB flash drive or external USB hard disk and connect it to the Raspberry Pi. You should be able to play them from XBMC.

Since the Raspberry Pi is likely to be sitting near your TV, you may find that your TV has a USB port that can provide sufficient current to run the Raspberry Pi so that you don't need a separate power supply.

A wireless keyboard and mouse are a good idea, since if you buy them as a pair, they will use a single USB port for the dongle and avoid wires trailing all over the place. You can also buy mini-keyboards with built-in trackpads that are useful in this situation.

A wired network connection is generally higher performance and better than a WiFi connection, but it is not always convenient to have the Pi near an Ethernet socket. If this is the case, then you can set up XBMC to use a WiFi dongle for a network connection.

Setting up WiFi when using XBMC and Raspbmc is easier than under Raspbian or Occidentalis (Recipe 2.5) since you can use the slick user interface provided by the media center software. To set up your wireless network, go to the Programs section of XBMC and then choose Raspbmc Settings (Figure 4-2).

Figure 4-2. Setting up WiFi in Raspbmc

Select the WiFi option and then enter your SSID (WiFi network name) and WiFi password.

See Also

The XBMC software has a user manual in the form of a wiki (*http://wiki.xbmc.org/*).

Raspbmc is not the only media center distribution. Two other popular distributions are:

- OpenElec (*http://www.openelec.tv*)
- XBian (*http://xbian.org*)

You can add an IR remote to Raspberry Pi to control XBMC (*http://bit.ly/17W0Ghl*).

4.2. Installing Office Software

Problem

You need to open word processor and spreadsheet documents on a Raspberry Pi.

Solution

Raspberry Pi is, after all, a Linux computer, so there are several office applications that you can install to use spreadsheets and word processor documents.

Programs for the Raspberry Pi are downloaded from the Internet, so you'll need an Internet connection.

Before you install any new software, it is a good idea to open a Terminal and run the following command:

```
$ sudo apt-get update
```

To install the AbiWord word processor, run this command:

```
$ sudo apt-get install abiword
```

You will be asked to confirm the installation by typing **Y**; after a minute or so, installation will be complete. If you look on your Start menu, you will see a new section called Office, where you will find AbiWord (Figure 4-3).

AbiWord will open *.doc*, *.docx*, and other common word processing document formats.

If you need to use a spreadsheet, Gnumeric is a good choice. Install it using the command:

```
$ sudo apt-get install gnumeric
```

Discussion

If you find that the office applications run a bit too slowly, try overclocking your Raspberry Pi (Recipe 1.14) to speed things up.

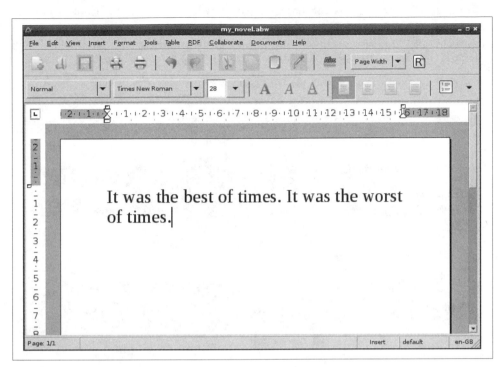

Figure 4-3. AbiWord on Raspberry Pi

See Also

There are efforts in progress to port other office software, such as Libre Office (a fork of Open Office). Check on the Internet for the latest news on office software ports.

See also Recipe 3.16 for information on using `apt-get`.

4.3. Installing other Browsers

Problem

You want to use a browser other than Midori.

Solution

You can use a number of browsers on the Raspberry Pi. The Pi is not a powerful computer, and modern browsers and web pages can put quite a load on it. This means that when you use a browser on a Raspberry Pi, there is always a trade-off between features and performance.

Chromium (Figure 4-4), as the name suggests, will be familiar to users of Google Chrome. It is full featured, but noticeably slow when you try to scroll up and down a busy web page. You can install Chromium using the following command, which will put a new link in the Internet section of your Start menu:

```
$ sudo apt-get install chromium-browser
```

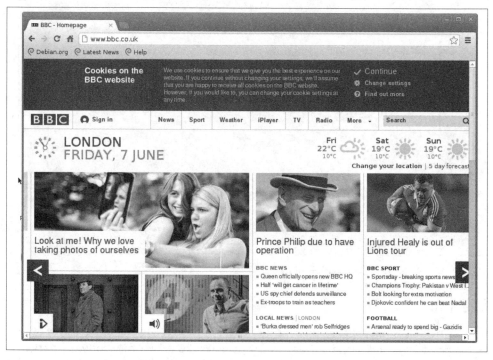

Figure 4-4. The Chromium browser

Another popular alternative to Midori is Iceweasel (Figure 4-5). This browser is based on Firefox and will perform faster than Chromium because it uses the mobile version of websites (when available), which will normally be much simpler HTML. You can download and install Iceweasel using this command:

```
$ sudo apt-get install iceweasel
```

Discussion

Since browsers require a certain amount of power for many websites, you really need a model B with 512 MB to avoid a frustratingly slow experience. It is probably also worth overclocking your Raspberry Pi (Recipe 1.14).

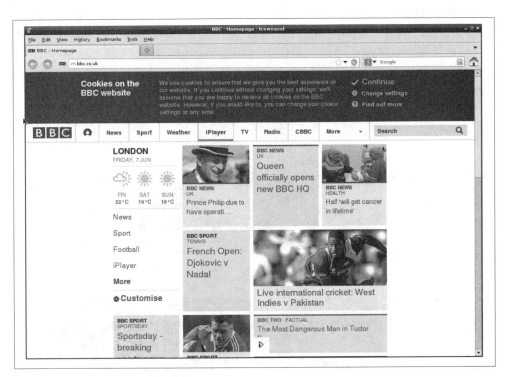

Figure 4-5. The Iceweasel browser

See Also

For more information on installing with `apt-get`, see Recipe 3.16.

4.4. Using the Pi Store

Problem

You want to install software and games using the Pi Store.

Solution

The Pi Store is the Raspberry Pi equivalent of Apple's App Store or Google's Play Store; it is a place from which you can download, install, and run all sorts of applications, both free and for sale.

If you want to use the Pi Store, the first step is to download the client program that will run on your Raspberry Pi and allow you to browse the apps available. To do this, open a Terminal session and issue the command:

```
$ sudo apt-get install pistore
```

Once installation is complete, you will find a new shortcut on your desktop for the Pi Store application (Figure 4-6).

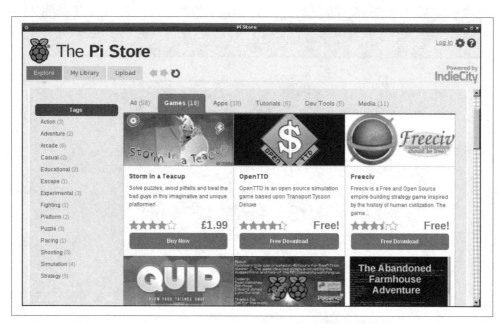

Figure 4-6. The Pi Store

The first time you try to download an app, you will be asked to register. After you register, the app will download and appear in the My Library tab. To run the app, just double-click it.

Discussion

This is a convenient way to browse for interesting programs to use on your Raspberry Pi. More and more apps are finding their way into the store.

See Also

See the official Pi Store website (*http://store.raspberrypi.com/*).

For more information on installing with `apt-get`, see Recipe 3.16.

4.5. Making a Webcam Server

Problem

You want to set up a Raspberry Pi as a webcam server.

Solution

Download the motion software. This will allow you to set up a Raspberry Pi with a USB webcam connected to it so you can connect to a web page and view the webcam.

To install the software, enter the following command in a Terminal window:

```
$ sudo apt-get install motion
```

Plug in your USB webcam and type **lsusb** to see if the webcam is connected:

```
$ lsusb
Bus 001 Device 001: ID 1d6b:0002 Linux Foundation 2.0 root hub
Bus 001 Device 002: ID 0424:9512 Standard Microsystems Corp.
Bus 001 Device 003: ID 0424:ec00 Standard Microsystems Corp.
Bus 001 Device 004: ID 3538:0059 Power Quotient International Co., Ltd
Bus 001 Device 006: ID eb1a:299f eMPIA Technology, Inc.
```

If it isn't obvious, try disconnecting it, run the command again, and see if one of the entries vanishes from the list. In this case, the final entry in the list is for the webcam.

There are now a few configuration changes that you need to make. First, edit the file */etc/motion/motion.conf* using the command:

```
$ sudo nano /etc/motion/motion.conf
```

This is quite a big configuration file. Right near the top of the file, you will find the line daemon off; change this to be daemon on.

The other change is much further down the file. You need to change webcam_localhost = on to webcam_localhost = off.

There is one other file that you need to change. Enter the command:

```
$ sudo nano /etc/default/motion
```

Change start_motion_daemon=no to start_motion_daemon=yes.

To get the web service running, issue the command:

```
$ sudo service motion start
```

You should now be able to open your web browser and view the webcam. To do this, you will need to know the IP address of your Raspberry Pi (Recipe 2.2).

From another computer on the same network, open a browser and go to the URL *http://192.168.1.16:8081/*. You will need to change the URL to match the IP address of your Raspberry Pi, but you need to keep the :8081 port number on the end of the URL.

If all goes well, you should see something like Figure 4-7.

Figure 4-7. A Raspberry Pi webcam

Discussion

The motion software is actually really powerful; there are many other settings you can tweak to change how your webcam works.

By default, the webcam will be viewable only from within your network. If you want your webcam to be visible to the whole Internet, you need to set up *port forwarding* on your home hub. This will require you to log in to the admin console for the hub, look for the port forwarding option, and enable it for port 8081 for the IP address of the Raspberry Pi.

You will then be able to view the webcam using the external IP address allocated by your ISP. This is usually shown on the from page of your admin console. But beware: unless you pay your ISP for a static IP address, this IP address is likely to change every time you restart your home hub modem.

See Also

There is thorough documentation on the motion website (*http://bit.ly/1dyfqkA*).

The Raspberry Pi has a plug-in camera module (Figure 1-18). At the time of writing, this is not compatible with motion, but it may be by the time you are reading this.

See also Recipe 3.16 for information on using `apt-get`.

4.6. Running a Vintage Game Console Emulator

Problem

You want to use a games emulator on a Raspberry Pi to make a vintage game console.

Solution

There are a number of emulators around for old game consoles from the 1980s. One of the most popular is Stella, an emulator for the Atari 2600 (Figure 4-8).

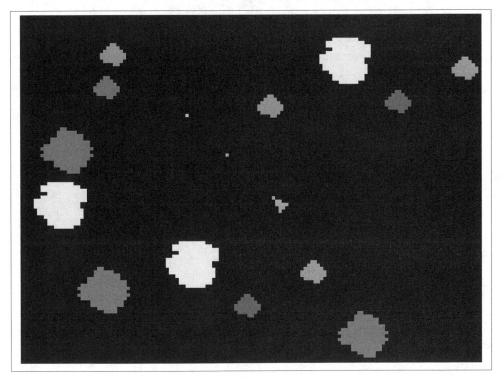

Figure 4-8. Asteroids on the Stella Atari 2600 emulator

 It is worth noting that even though these games are ancient, they are still owned by someone. The ROM image files that you need to play the games on an emulator like Stella—although easy to find on the Internet—are not necessarily yours to take. So, please stick to the law.

To install Stella, simply enter the following command in a Terminal window:

```
$ sudo apt-get install stella
```

Once it's installed, you will find a new program in Games on your Start menu. Do not run it yet—you need to obtain the ROM image for a game.

If you are outside of the US and own the actual game, you are entitled to take a copy of the ROM image for backup purposes in many, but not all, countries. You can also find ROM images for some games that have been released without any license restrictions.

Now that you have the ROM image for the game you wish to play, create a folder called *roms* and place the ROM image in that folder. Then start Stella.

Click the image file to start a game. By default, the cursor keys are mapped to the joystick controls and the space bar is fire. You can change a number of settings for the emulator —for example, you almost certainly want to go to Video Settings and set it to fullscreen mode.

You can also use Input Settings and the Emul Events tab to map controller buttons to your keyboard.

Discussion

The emulator uses a surprisingly large amount of the Raspberry Pi's meager resources, so you may find that you need to overclock your Pi for best performance (Recipe 1.14).

If you search the Internet, you will find that many people have taken this basic setup and added a retro USB controller, like the widely available and quite low-cost Nintendo Retrolink USB Super SNES Classic Controller, and built the Pi and a monitor into a big arcade-style housing.

See Also

There are a number of other console emulators available—at various stages of maturity, and thus reliability—on the Raspberry Pi. One to consider is Mame (*http://www.mame dev.org/*), which emulates a number of different game platforms.

For more information on installing with `apt-get`, see Recipe 3.16.

4.7. Running Minecraft

Problem

You want to run the popular game *Minecraft* on your Raspberry Pi.

Solution

Mojang, the original developers of *Minecraft*, have ported it to the Raspberry Pi.

To install the software, you need to be using Raspbian (Recipe 1.4). Enter the following commands to download and install *Minecraft* (Figure 4-9):

```
$ wget https://s3.amazonaws.com/assets.minecraft.net/pi/minecraft-pi-0.1.1.tar.gz
$ tar -zxvf minecraft-pi-0.1.1.tar.gz
$ cd mcpi
$ ./minecraft-pi
```

Figure 4-9. Minecraft on Raspberry Pi

Discussion

To get *Minecraft* to fit on the Raspberry Pi, the developers made some shortcuts in the graphics code. This means that you can only play the game directly on the Raspberry Pi with the keyboard, mouse, and monitor connected directly to it. It will not work over a remote connection.

See Also

Find out more about the Raspberry Pi port of *Minecraft* (*http://pi.minecraft.net/*).

4.8. Running Open Arena

Problem

You want to run the *Quake* derivative *Open Arena* on your Raspberry Pi.

Solution

Download and run *Open Arena* (Figure 4-10) from the Pi Store.

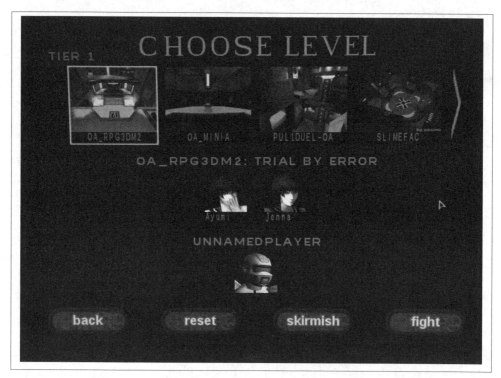

Figure 4-10. Open Arena on Raspberry Pi

Discussion

Not surprisingly, you will find *Open Arena* in the Games section of the Pi Store. Since the game is quite violent and bloody, it comes under a plain wrapper.

See Also

Find out more about *Open Arena* (*http://www.openarena.ws/*).

For more information on using the Pi Store, see Recipe 4.4.

4.9. Raspberry Pi Radio Transmitter

 Be sure to check out the accompanying video for this recipe at *http://razzpisampler.oreilly.com*.

Problem

You want to convert your Raspberry Pi into a high-powered FM transmitter that will send a radio signal to a normal FM radio receiver (Figure 4-11).

Solution

The clever folks at Imperial College London have created some C code and a Python wrapper that allow you to do just this. The download even includes the theme from *Star Wars* to play as a sample.

All you need is a short length of wire attached to GPIO pin 4. A female-to-male header lead will work just fine for this. In fact, it should work with the radio right next to your Pi without any kind of antenna—such is the strength of the transmission.

The first step is to install the *pifm* library using the following commands:

```
$ mkdir pifm
$ cd pifm
$ wget http://www.icrobotics.co.uk/wiki/images/c/c3/Pifm.tar.gz
$ tar -xzf Pifm.tar.gz
```

Next, find yourself an FM radio receiver and tune it to 103.0 MHz. If this frequency is already occupied by some other transmission, pick another frequency and make note of it.

Now run the following command (changing the final parameter from 103.0 to a different frequency if you had to change frequency):

```
sudo ./pifm sound.wav 103.0
```

If all is well, you should hear the stirring tones of the *Star Wars* theme.

Figure 4-11. Raspberry Pi as an FM transmitter

Discussion

You need to know that this project may not be legal in your country. The power output is higher than that of FM transmitters used with MP3 players.

You can play other *.wav* files, but they must be 16-bit 44.1kHz mono.

The code also includes a Python library that you can use within your own Python programs. So, you could write yourself a user interface to allow the selection and playing of tunes.

The following fragment of code illustrates the use of the Python interface:

```
pi@raspberrypi ~/pifm $ sudo python
Python 2.7.3 (default, Jan 13 2013, 11:20:46)
[GCC 4.6.3] on linux2
Type "help", "copyright", "credits" or "license" for more information.
>>> import PiFm
>>> PiFm.play_sound("sound.wav")
```

Were you to put a Raspberry Pi in your vehicle, this would be a great way of outputting sound through the vehicle's audio system.

See Also

This description is based on the original post from Imperial College London (*http://bit.ly/18AcT5u*).

4.10. Running GIMP

Problem

You want to edit an image.

Solution

Download and run the GNU image manipulation program (GIMP; see Figure 4-12).

Figure 4-12. GIMP on the Raspberry Pi

To install GIMP, open a Terminal session and type the following command:

```
$ sudo apt-get install gimp
```

Once GIMP is installed, you'll find a new entry in your Start menu under the Graphics heading called GNU Image Manipulation Program.

Discussion

Despite being very hungry for memory and processor power, GIMP is quite usable on a Raspberry Pi model B.

See Also

Find out more from the GIMP website (*http://www.gimp.org/*).

GIMP has a lot of features and is a very sophisticated image editing program, so it does take a little learning. You'll find an online manual for the software at the GIMP website, under the Documentation tab.

For more information on installing with `apt-get`, see Recipe 3.16.

4.11. Internet Radio

Problem

You want to be able to play Internet radio on your Raspberry Pi.

Solution

Install the VLC media player by running the following command:

```
sudo apt-get install vlc
```

Once it's installed, you can find VLC in the Sound & Video section of your Start menu.

Run the program and select the Open Network Stream option on the Media menu. This will open a dialog box (see Figure 4-13) where you can enter the URL of the Internet radio station that you wish to play.

You will need to plug headphones or amplified speakers into the audio socket on the Raspberry Pi.

Discussion

VLC can also be run from the command line:

```
$ vlc http://www.a-1radio.com/listen.pls -I dummy
```

VLC will probably produce a series of error messages, but then play the audio just fine.

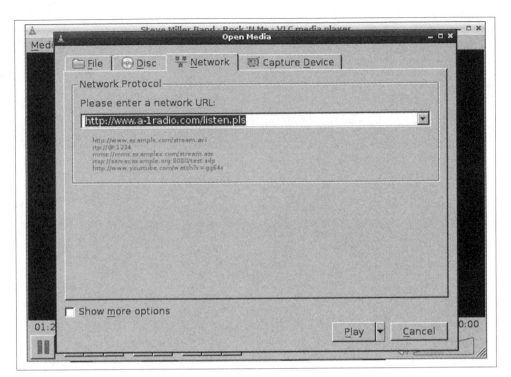

Figure 4-13. VLC on Raspberry Pi

See Also

This recipe borrows heavily from this tutorial (*http://www.jan-holst.dk/pi-radio/pi-radio.html*), where Jan Holst takes things a step further and adds radio-style controls to the project.

For UK readers, you can find a list of the BBC radio stream URLs (*http://forum.chumby.com/viewtopic.php?id=7054*).

Python Basics

5.0. Introduction

Although there are many languages that can be used to program the Raspberry Pi, Python is the most popular. In fact, the *Pi* in *Raspberry Pi* is inspired by the word Python.

In this chapter, you will find a host of recipes to help you get programming with Raspberry Pi.

5.1. Deciding Between Python 2 and Python 3

Problem

You need to use Python but are unsure which version to use.

Solution

Use both. Use Python 3 until you face a problem that is best solved by reverting to version 2.

Discussion

Although the latest version of Python is Python 3 and has been for years, you will find that a lot of people stick to Python 2. In fact, in the Raspbian and Occidentalis distributions, both versions are supplied and version 2 is just called Python, whereas version 3 is called Python 3. You even run Python version 3 using the command python3. The examples in this book are written for Python 3 unless otherwise stated. Most will run on both Python 2 and Python 3 without modification.

This reluctance on the part of the Python community to ditch the old Python 2 is largely because Python 3 introduced some changes that broke compatibility with version 2.

This means that some of the huge body of third-party libraries that have been developed for Python 2 won't work under Python 3.

My strategy is to write in Python 3 wherever possible, and revert to Python 2 when I have to because of compatibility problems.

See Also

For a good summary of the Python 2 versus Python 3 debate, see the Python wiki (*http://wiki.python.org/moin/Python2orPython3*).

5.2. Editing Python Programs with IDLE

Problem

You are not sure what to use to write your Python programs.

Solution

The common Raspberry Pi distributions come with the IDLE Python development tool in both the Python and Python 3 versions. If you are using Raspbian or Occidentalis, you will find shortcuts to both versions of IDLE on your Raspberry Pi desktop.

Discussion

IDLE and IDLE 3 appear identical; the only difference is the version of Python that they use, so open IDLE 3 (Figure 5-1).

```
Python Shell                                                    _ □ x
File  Edit  Shell  Debug  Options  Windows  Help
Python 2.7.3 (default, Jan 13 2013, 11:20:46)
[GCC 4.6.3] on linux2
Type "copyright", "credits" or "license()" for more information.
>>> |

                                                            Ln: 4 Col: 4
```

Figure 5-1. IDLE Python console

The window that opens is labeled Python Shell. This is an interactive area where you can type **Python** and see the results immediately. So try typing the following into the shell next to the >>> prompt:

```
>>> 2 + 2
4
>>>
```

Python has evaluated the expression 2 + 2 and given us the answer of 4.

The Python Shell is fine for trying things out, but if you want to write programs, then you need to use an editor. To open a new file to edit using IDLE, select New Window from the File menu (Figure 5-2).

Figure 5-2. IDLE editor

You can now enter your program into the editor window. Just to try things, paste the following text into the editor, save the file as *count.py*, and then select Run Module from the Run menu. You can see the results of the program running in the Python console:

```
for i in range(1, 10):
    print(i)
```

Python is unusual for a programming language in that indentation is a fundamental part of the language. Whereas many C-based languages use { and } to delimit a block of code, Python uses the indentation level. So, in the preceding example, Python knows that print is to be invoked repeatedly as part of the for loop because it is indented out four spaces.

The convention used in this book is to use four spaces for a level of indentation.

 When you're starting out in Python, it's not uncommon to see an error such as IndentationError: unexpected indent, which means that somewhere things are not indented correctly. If everything appears to line up, double-check that none of the indents contain Tab characters. Python treats tabs differently.

Notice how IDLE uses color coding to highlight the structure of the program.

See Also

Many of the recipes in this chapter use IDLE to edit Python examples.

As well as using IDLE to edit and run Python files, you can also edit files in nano (Recipe 3.6) and then run them from a Terminal session (Recipe 5.4).

5.3. Using the Python Console

Problem

You want to enter Python commands without writing a whole program.

Solution

Use the Python console, either within IDLE (Recipe 5.2) or in a Terminal session.

Discussion

To start a Python 2 console in a Terminal window, just type the command **python**; for a Python 3 console, enter the command **python3**.

The >>> prompt indicates that you can type Python commands. If you need to type multiline commands, then the console will automatically provide a continuation line indicated by three dots. You still need to indent any such lines, by four spaces, as shown in the following session:

```
>>> for i in range(1, 10):
...     print(i)
...
1
2
3
4
5
6
7
8
```

```
9
>>>
```

You will need to press Enter twice after your last command for the console to recognize the end of the indented block and run the code.

The Python console also provides a command history so that you can move back and forth through your previous commands using the up and down keys.

See Also

If you have more than a couple of lines that you want to type in, then chances are you would be better off using IDLE (Recipe 5.2) to edit and run a file.

5.4. Running Python Programs from the Terminal

Problem

Running programs from within IDLE is fine, but sometimes you want to run a Python program from a Terminal window.

Solution

Use the python or python3 command in a Terminal, followed by the filename containing the program you want to run.

Discussion

To run a Python 2 program from the command line, use a command like this:

```
$ python myprogram.py
```

If you want to run the program using Python 3, then change the command python to python3. The Python program that you wish to run should be in a file with the extension *.py*.

You can run most Python programs as a normal user; however, there are some, especially those that use the GPIO port, that you need to run as super user. If this is the case for your program, prefix the command with sudo:

```
$ sudo python myprogram.py
```

See Also

Recipe 3.21 allows you to run a Python program as a timed event.

To automatically run a program at startup, see Recipe 3.20.

5.5. Variables

Problem

You want to give a value a name.

Solution

Assign a value to a name using =.

Discussion

In Python, you don't have to declare the type of a variable, you can just assign it a value as shown in the following examples:

```
a = 123
b = 12.34
c = "Hello"
d = 'Hello'
e = True
```

You can define character string constants using either single or double quotes. The logical constants in Python are True and False and are case-sensitive.

By convention, variable names begin with a lowercase letter and if the variable name consists of more than one word, the words are joined together with an underscore character. It is always a good idea to give your variables descriptive names.

Some examples of valid variable names are x, total, and number_of_chars.

See Also

Variables can also be assigned a value that is a list (Recipe 6.1) or dictionary (Recipe 6.12).

For more information on arithmetic with variables, see Recipe 5.8.

5.6. Displaying Output

Problem

You want to see the value of a variable.

Solution

Use the print command. You can try the following example in the Python console (Recipe 5.3):

```
>>> x = 10
>>> print(x)
10
>>>
```

Discussion

In Python 2, you can use the `print` command without brackets. However, this is not true in Python 3, so for compatibility with both versions of Python, use brackets around the value you are printing.

See Also

To read user input, see Recipe 5.7.

5.7. Reading User Input

Problem

You want to prompt the user to enter a value.

Solution

Use the `input` (Python 3) or `raw_input` (Python 2) command. You can try the following example in the Python 3 console (Recipe 5.3):

```
>>> x = input("Enter Value:")
Enter Value:23
>>> print(x)
23
>>>
```

Discussion

In Python 2, `raw_input` must be substituted for `input` in the preceding example.

Python 2 also has an `input` function, but this validates the input and attempts to convert it into a Python value of the appropriate type, whereas `raw_input` does the same thing as `input` in Python 3 and just returns a string.

See Also

Find more information on Python 2 input (*http://docs.python.org/2/library/func tions.html#input*).

5.8. Arithmetic

Problem

You want to do arithmetic in Python.

Solution

Use the +, -, *, and / operators.

Discussion

The most common operators for arithmetic are +, -, *, and /, which are, respectively, add, subtract, multiply, and divide.

You can also group parts of the expression together with parentheses, as shown in the following example, which, given a temperature in degrees Celsius, converts it to degrees Fahrenheit:

```
>>> tempC = input("Enter temp in C: ")
Enter temp in C: 20
>>> tempF = (int(tempC) * 9) / 5 + 32
>>> print(tempF)
68.0
>>>
```

Other arithmetic operators include % (modulo remainder) and ** (raise to the power of). For example, to raise 2 to the power of 8, you would write the following:

```
>>> 2 ** 8
256
```

See Also

See Recipe 5.7 on using the input command, and Recipe 5.7 on converting the string value from input to a number.

The Math library has many useful math functions (*http://docs.python.org/3.0/library/math.html*) that you can use.

5.9. Creating Strings

Problem

You want to create a string variable.

Solution

Use the assignment operator and a string constant to create a new string. You can use either double or single quotation marks around the string, but they must match.

For example:

```
>>> s = "abc def"
>>> print(s)
abc def
>>>
```

Discussion

If you need to include double or single quotes inside a string, then pick the type of quotes that you don't want to use inside the string as the beginning and end markers of the string. For example:

```
>>> s = "Isn't it warm?"
>>> print(s)
Isn't it warm?
>>>
```

Sometimes you'll need to include special characters such as tab or newline inside your string. This requires the use of what are called *escape characters*. To include a tab, use \t, and for a newline, use \n. For example:

```
>>> s = "name\tage\nMatt\t14"
>>> print(s)
name    age
Matt    14
>>>
```

See Also

For a full list of escape characters, see the Python Reference Manual (*http://bit.ly/17Xxuqf*).

5.10. Concatenating (Joining) Strings

Problem

You want to join a number of strings together.

Solution

Use the + (concatenate) operator.

For example:

```
>>> s1 = "abc"
>>> s2 = "def"
>>> s = s1 + s2
>>> print(s)
abcdef
>>>
```

Discussion

In many languages, you can have a chain of values to concatenate, some of which are strings and some of which are other types such as numbers; numbers will automatically be converted into strings during the concatenation. This is not the case in Python, and if you try the following command, you will get an error:

```
>>> "abc" + 23
Traceback (most recent call last):
  File "<stdin>", line 1, in <module>
TypeError: Can't convert 'int' object to str implicitly
```

You need to convert each component that you want to concatenate to a string before concatenating, as shown in this example:

```
>>> "abc" + str(23)
'abc23'
>>>
```

See Also

See Recipe 5.11 for more information about converting numbers to strings using the str function.

5.11. Converting Numbers to Strings

Problem

You want to convert a number into a string.

Solution

Use the str Python function. For example:

```
>>> str(123)
'123'
>>>
```

Discussion

A common reason for wanting to convert a number into a string is so you can then concatenate it with another string (Recipe 5.10).

See Also

For the reverse operation of turning a string into a number, see Recipe 5.12.

5.12. Converting Strings to Numbers

Problem

You want to convert a string into a number.

Solution

Use the `int` or `float` Python function.

For example, to convert -123 into a number, you could use:

```
>>> int("-123")
-123
>>>
```

This will work on both positive and negative whole numbers.

To convert a floating-point number, use `float` instead of `int`:

```
>>> float("00123.45")
123.45
>>>
```

Discussion

Both `int` and `float` will handle leading zeros correctly and are tolerant of any spaces or other whitespace characters around the number.

You can also use `int` to convert a string representing a number in a number base other than the default of 10 by supplying the number base as the second argument. The following example converts the string representation of binary 1001 into a number:

```
>>> int("1001", 2)
9
>>>
```

This second example converts the hexadecimal number AFF0 to an integer:

```
>>> int("AFF0", 16)
45040
>>>
```

See Also

For the reverse operation of turning a number into a string, see Recipe 5.11.

5.13. Find the Length of a String

Problem

You need to know how many characters there are in a string.

Solution

Use the len Python function.

Discussion

For example, to find the length of the string abcdef, you would use:

```
>>> len("abcdef")
6
>>>
```

See Also

The len command also works on arrays (Recipe 6.3).

5.14. Find the Position of One String Inside Another

Problem

You need to find the position of one string within another.

Solution

Use the find Python function.

For example, to find the starting position of the string def within the string abcdef
ghi, you would use:

```
>>> s = "abcdefghi"
>>> s.find("def")
3
>>>
```

Note that the character positions start at 0, so a position of 3 means the fourth character in the string.

Discussion

If the string you're looking for doesn't exist in the string being searched, then find returns the value -1.

See Also

The replace function is used to both find and then replace all occurrences of a string (Recipe 5.16).

5.15. Extracting Part of a String

Problem

You want to cut out a section of a string between certain character positions.

Solution

Use the Python [:] notation.

For example, to cut out a section from the second character to the fifth character of the string abcdefghi, you would use:

```
>>> s = "abcdefghi"
>>> s[1:5]
'bcde'
>>>
```

Note that the character positions start at 0, so a position of 1 means the second character in the string and 5 means the sixth, but the character range is exclusive at the high end, so the letter *f* is not included in this example.

Discussion

The [:] notation is actually quite powerful. You can omit either argument, in which case, the start or end of the string is assumed as appropriate. For example:

```
>>> s[:5]
'abcde'
>>>
```

and

```
>>> s = "abcdefghi"
>>> s[3:]
```

```
'defghi'
>>>
```

You can also use negative indices to count back from the end of the string. This can be useful in situations such as when you want to find the three-letter extension of a file, as in the following example:

```
>>> "myfile.txt"[-3:]
'txt'
```

See Also

Recipe 5.10 describes joining strings together rather than splitting them.

Recipe 6.10 uses the same syntax but with lists.

5.16. Replacing One String of Characters with Another Inside a String

Problem

You want to replace all occurrences of a string within another string.

Solution

Use the `replace` function.

For example, to replace all occurrences of X with `times`, you would use:

```
>>> s = "It was the best of X. It was the worst of X"
>>> s.replace("X", "times")
'It was the best of times. It was the worst of times'
>>>
```

Discussion

The string being searched for must match exactly; that is, the search is case-sensitive and will include spaces.

See Also

See Recipe 5.14 for searching a string without performing a replacement.

5.17. Converting a String to Upper- or Lowercase

Problem

You want to convert all the characters in a string to upper- or lowercase letters.

Solution

Use the upper or lower function as appropriate.

For example, to convert aBcDe to uppercase, you would use:

```
>>> "aBcDe".upper()
'ABCDE'
>>>
```

To convert it to lowercase, you would use:

```
>>> "aBcDe".lower()
'abcde'
>>>
```

Discussion

In common with most functions that manipulate a string in some way, upper and lower do not actually modify the string, but rather return a modified copy of the string.

For example, the following code returns a copy of the string s, but note how the original string is unchanged.

```
>>> s = "aBcDe"
>>> s.upper()
'ABCDE'
>>> s
'aBcDe'
>>>
```

If you want to change the value of s to be all uppercase, then you need to do the following:

```
>>> s = "aBcDe"
>>> s = s.upper()
>>> s
'ABCDE'
>>>
```

See Also

See Recipe 5.16 for replacing text within strings.

5.18. Running Commands Conditionally

Problem

You want to run some Python commands only when some condition is true.

Solution

Use the Python `if` command.

The following example will print the message `x is big` only if `x` has a value greater than 100:

```
>>> x = 101
>>> if x > 100:
...     print("x is big")
...
x is big
```

Discussion

After the `if` keyword, there is a *condition*. This condition often, but not always, compares two values and gives an answer that is either `True` or `False`. If it is `True`, then the subsequent indented lines will all be executed.

It is quite common to want to do one thing if a condition is `True` and something different if the answer is `False`. In this case, the `else` command is used with `if`, as shown in this example:

```
x = 101
if x > 100:
    print("x is big")
else:
    print("x is small")

print("This will always print")
```

You can also chain together a long series of `elif` conditions. If any one of the conditions succeeds, then that block of code is executed, but none of the other conditions are tried. For example:

```
x = 90
if x > 100:
    print("x is big")
elif x < 10:
    print("x is small")
else:
    print("x is medium")
```

This example will print `x is medium`.

See Also

See Recipe 5.18 for more information on different types of comparisons you can make.

5.19. Comparing Values

Problem

You want to compare values with each other.

Solution

Use one of the comparison operators: <, >, <=], >=, ==, or !=.

Discussion

You use the < (less than) and > (greater than) operators in Recipe 6.15. Here's the full set of comparison operators:

<
 Less than

>
 Greater than

<=
 Less than or equal to

>=
 Greater than or equal to

==
 Exactly equal to

!=
 Not equal to

Some people prefer to use the <> operator in place of !=. Both work the same.

You can test out these commands using the Python console (Recipe 5.3), as shown in the following exchange:

```
>>> 1 != 2
True
>>> 1 != 1
False
>>> 10 >= 10
True
>>> 10 >= 11
```

```
False
>>> 10 == 10
True
>>>
```

A common mistake is to use = (set a value) instead of == (double equals) in comparisons. This can be difficult to spot because if one half of the comparison is a variable, it is perfectly legal syntax and will run, but it will not produce the result you were expecting.

As well as comparing numbers, you can also compare strings using these comparison operators, for example:

```
>>> 'aa' < 'ab'
True
>>> 'aaa' < 'aa'
False
```

The strings are compared lexicographically—that is, in the order that you would find them in a dictionary.

This is not quite correct as, for each letter, the uppercase version of the letter is considered less than the lowercase equivalent.

See Also

See also Recipe 6.15.

5.20. Logical Operators

Problem

You need to specify a complex condition in an if statement.

Solution

Use one of the logical operators: and, or, and not.

Discussion

As an example, you may want to check whether a variable x has a value between 10 and 20. For that, you would use the and operator:

```
>>> x = 17
>>> if x >= 10 and x <= 20:
...     print('x is in the middle')
...
x is in the middle
```

You can combine as many and and or statements as you need and also use brackets to group them if the expressions become complicated.

See Also

See Recipes 5.18 and 6.15.

5.21. Repeating Instructions an Exact Number of Times

Problem

You need to repeat some program code an exact number of times.

Solution

Use the Python for command and iterate over a range.

For example, to repeat a command 10 times, use the following example:

```
>>> for i in range(1, 11):
...     print(i)
...
1
2
3
4
5
6
7
8
9
10
>>>
```

Discussion

The second parameter in the range command is exclusive; that is, to count up to 10, you must specify a value of 11.

See Also

If the condition for stopping the loop is more complicated than simply repeating a number of times, see Recipe 5.21.

If you are trying to repeat commands for each element of a list or dictionary, see Recipes 6.7 or 6.15, respectively.

5.22. Repeating Instructions Until Some Condition Changes

Problem

You need to repeat some program code until something changes.

Solution

Use the Python `while` statement. The `while` statement repeats its nested commands until its condition becomes false. The following example will stay in the loop until the user enters *X* for exit:

```
>>> answer = ''
>>> while answer != 'X':
...     answer = input('Enter command:')
...
Enter command:A
Enter command:B
Enter command:X
>>>
```

Discussion

Note that the preceding example uses the `input` command as it works in Python 3. To run the example in Python 2, substitute the command `raw_input` for `input`.

See Also

If you just want to repeat some commands a number of times, see Recipe 5.20.

If you are trying to repeat commands for each element of a list or dictionary, then see Recipes 6.7 or 6.15, respectively.

5.23. Breaking Out of a Loop

Problem

You are in a loop and need to exit the loop if some condition occurs.

Solution

Use the Python `break` statement to exit either a `while` or `for` loop.

The following example behaves in exactly the same way as the example code in Recipe 5.22:

```
>>> while True:
...     answer = input('Enter command:')
...     if answer == 'X':
...             break
...
Enter command:A
Enter command:B
Enter command:X
>>>
```

Discussion

Note that this example uses the input command as it works in Python 3. To run the example in Python 2, substitute the command raw_input for input.

This example behaves in exactly the same way as the example of Recipe 5.21. However, in this case the condition for the while loop is just True, so the loop will never end unless we use break to exit the loop when the user enters *X*.

See Also

You can also leave a while loop using its condition; see Recipe 5.21.

5.24. Defining a Function in Python

Problem

You want to avoid repeating the same code over and over in a program.

Solution

Create a function that groups together lines of code, allowing it to be called from multiple places.

Creating and then calling function in Python is illustrated in the following example:

```
def count_to_10():
    for i in range(1, 11):
        print(i)

count_to_10()
```

In this example, we have defined a function using the def command that will print out the numbers between 1 and 10 whenever it is called:

```
count_to_10()
```

Discussion

The conventions for naming functions are the same as for variables in Recipe 5.5; that is, they should start with a lowercase letter, and if the name consists of more than one word, the words should be separated by underscores.

The example function is a little inflexible because it can only count to 10. If we wanted to make it more flexible—so it could count up to any number—then we can include the maximum number as a *parameter* to the function, as this example illustrates:

```
def count_to_n(n):
    for i in range(1, n + 1):
        print(i)

count_to_n(5)
```

The parameter name *n* is included inside the parentheses and then used inside the range command, but not before 1 is added to it.

Using a parameter for the number we want to count up to means that if we usually count to 10, but sometimes count to a different number, we will always have to specify the number. You can, however, specify a default value for a parameter, and hence have the best of both worlds, as shown in this example:

```
def count_to_n(n=10):
    for i in range(1, n + 1):
        print(i)

count_to_n()
```

This will now count to 10 unless a different number is specified when you call the function.

If your function needs more than one parameter, perhaps to count between two numbers, then the parameters are separated by commas:

```
def count(from_num=1, to_num=10):
    for i in range(from_num, to_num + 1):
            print(i)

count()
count(5)
count(5, 10)
```

All these examples are functions that do not return any value, they just do something. If you need a function to return a value, you need to use the return command.

The following function takes a string as an argument and adds the word *please* to the end of the string.

```
def make_polite(sentence):
    return sentence + " please"

print(make_polite("Pass the cheese"))
```

When a function returns a value, you can assign the result to a variable, or as in this example, print out the result.

See Also

To return more than one value from a function, see Recipe 7.3.

Python Lists and Dictionaries

6.0. Introduction

In Chapter 5, we looked at the basics of the Python language. In this chapter, we look at two key Python data structures, lists and dictionaries.

6.1. Creating a List

Problem

You want to use a variable to hold a series of values rather than just one value.

Solution

Use a list. In Python, a list is a collection of values stored in order so that you can access them by position.

You create a list using [and] to contain its initial contents:

```
>>> a = [34, 'Fred', 12, False, 72.3]
>>>
```

Unlike more rigid arrays in languages like C, you don't need to specify the size of a list in Python when you declare it. You can also change the number of elements in the list any time you like.

Discussion

As this example illustrates, the items in a list do not have to be all of the same type, although they often are.

If you want to create an empty list that you can add items to later, you can write:

```
>>> a = []
>>>
```

See Also

All the recipes between Recipes 6.1 and 6.11 involve the use of lists.

6.2. Accessing Elements of a List

Problem

You want to find individual elements of a list or change them.

Solution

Use the [] notation to access elements of a list by their position in the list. For example:

```
>>> a = [34, 'Fred', 12, False, 72.3]
>>> a[1]
'Fred'
```

Discussion

The list positions (indices) start at 0 for the first element.

As well as using the [] notation to read values out of a list, you can also use it to change values at a certain position. For example:

```
>>> a = [34, 'Fred', 12, False, 72.3]
>>> a[1] = 777
>>> a
[34, 777, 12, False, 72.3]
```

If you try to change (or, for that matter, read) an element of a list using an index that is too large, you will get an "Index out of range" error:

```
>>> a[50] = 777
Traceback (most recent call last):
  File "<stdin>", line 1, in <module>
IndexError: list assignment index out of range
>>>
```

See Also

All the recipes between Recipes 6.1 and 6.11 involve the use of lists.

6.3. Find the Length of a List

Problem

You need to know how many elements there are in a list.

Solution

Use the `len` Python function. For example:

```
>>> a = [34, 'Fred', 12, False, 72.3]
>>> len(a)
5
```

Discussion

The `len` command also works on strings (Recipe 5.13).

See Also

All the recipes between Recipes 6.1 and 6.11 involve the use of lists.

6.4. Adding Elements to a List

Problem

You need to add an item to a list.

Solution

Use the `append`, `insert`, or `extend` Python functions.

To add a single item to the end of a list, use `append`:

```
>>> a = [34, 'Fred', 12, False, 72.3]
>>> a.append("new")
>>> a
[34, 'Fred', 12, False, 72.3, 'new']
```

Discussion

Sometimes you don't want to add the new elements to the end of a list, but rather you want to insert them at a certain position in the list. For this, use the `insert` command. The first argument is the index where the item should be inserted, and the second argument is the item to be inserted:

```
>>> a.insert(2, "new2")
>>> a
[34, 'Fred', 'new2', 12, False, 72.3]
```

Note how all the elements after the newly inserted element are shuffled up one position.

Both `append` and `insert` add only one element to a list. The `extend` function adds all the elements of one list to the end of another:

```
>>> a = [34, 'Fred', 12, False, 72.3]
>>> b = [74, 75]
>>> a.extend(b)
>>> a
[34, 'Fred', 12, False, 72.3, 74, 75]
```

See Also

All the recipes between Recipes 6.1 and 6.11 involve the use of lists.

6.5. Removing Elements from a List

Problem

You need to remove an item from a list.

Solution

Use the `pop` Python function.

The command `pop` with no parameters removes the last element of a list:

```
>>> a = [34, 'Fred', 12, False, 72.3]
>>> a.pop()
72.3
>>> a
[34, 'Fred', 12, False]
```

Discussion

Notice that `pop` returns the value removed from the list.

To remove an item in a position rather than the last element, use `pop` with a parameter of the position from which the item will be removed:

```
>>> a = [34, 'Fred', 12, False, 72.3]
>>> a.pop(0)
34
```

If you use an index position that is beyond the end of the list, you will get an "Index out of range" exception.

See Also

All the recipes between Recipes 6.1 and 6.11 involve the use of lists.

6.6. Creating a List by Parsing a String

Problem

You need to convert a string of words separated by some character into an array of strings with each string in the array being one of the words.

Solution

Use the `split` Python string function.

The command `split` with no parameters separates the words out of a string into individual elements of an array:

```
>>> "abc def ghi".split()
['abc', 'def', 'ghi']
```

If you supply `split` with a parameter, then it will split the string using the parameter as a separator. For example:

```
>>> "abc--de--ghi".split('--')
['abc', 'de', 'ghi']
```

Discussion

This command can be very useful when you are, say, importing data from a file. The `split` command can optionally take an argument that is the string to use as a delimiter when you are splitting the string. So, if you were to use commas as a separator, you could split the string as follows:

```
>>> "abc,def,ghi".split(',')
['abc', 'def', 'ghi']
```

See Also

All the recipes between Recipes 6.1 and 6.11 involve the use of lists.

6.7. Iterating over a List

Problem

You need to apply some lines of code to each item in a list in turn.

Solution

Use the `for` Python language command:

```
>>> a = [34, 'Fred', 12, False, 72.3]
>>> for x in a:
...     print(x)
...
34
Fred
12
False
72.3
>>>
```

Discussion

The `for` keyword is immediately followed by a variable name (in this case, *x*). This is called the loop variable and will be set to each of the elements of the list specified after *in*.

The indented lines that follow will be executed one time for each element in the list. Each time around the loop, *x* will be given the value of the element in the list at that position. You can then use *x* to print out the value, as shown in the previous example.

See Also

All the recipes between Recipes 6.1 and 6.11 involve the use of lists.

6.8. Enumerating a List

Problem

You need to run some lines of code to each item in a list in turn, but you also need to know the index position of each item.

Solution

Use the `for` Python language command along with the `enumerate` command.

```
>>> a = [34, 'Fred', 12, False, 72.3]
>>> for (i, x) in enumerate(a):
...     print(i, x)
...
(0, 34)
(1, 'Fred')
(2, 12)
(3, False)
```

```
(4, 72.3)
>>>
```

Discussion

It's quite common to need to know the position of something in the list while enumerating each of the values. An alternative method is to simply count with an index variable and then access the value using the [] syntax:

```
>>> a = [34, 'Fred', 12, False, 72.3]
>>> for i in range(len(a)):
...     print(i, a[i])
...
(0, 34)
(1, 'Fred')
(2, 12)
(3, False)
(4, 72.3)
>>>
```

See Also

All the recipes between Recipes 6.1 and 6.11 involve the use of lists.

See Recipe 6.7 to iterate over a list without needing to know each item's index position.

6.9. Sorting a List

Problem

You need to sort the elements of a list.

Solution

Use the sort Python language command:

```
>>> a = ["it", "was", "the", "best", "of", "times"]
>>> a.sort()
>>> a
['best', 'it', 'of', 'the', 'times', 'was']
```

Discussion

When you sort a list, you'll actually be modifying it rather than returning a sorted copy of the original list. This means that if you also need the original list, you need to use the copy command in the standard library to make a copy of the original list before sorting it:

```
>>> import copy
>>> a = ["it", "was", "the", "best", "of", "times"]
>>> b = copy.copy(a)
>>> b.sort()
>>> a
['it', 'was', 'the', 'best', 'of', 'times']
>>> b
['best', 'it', 'of', 'the', 'times', 'was']
>>>
```

See Also

All the recipes between Recipes 6.1 and 6.11 involve the use of lists.

6.10. Cutting Up a List

Problem

You need to make a sublist of a list using a range of the original list's elements.

Solution

Use the [:] Python language construction. The following example returns a list containing the elements of the original list from index position 1 to index position 2 (the number after the : is exclusive):

```
>>> l = ["a", "b", "c", "d"]
>>> l[1:3]
['b', 'c']
```

Note that the character positions start at 0, so a position of 1 means the second character in the string and 5 means the sixth, but the character range is exclusive at the high end, so the letter *d* is not included in this example.

Discussion

The [:] notation is actually quite powerful. You can omit either argument, in which case the start or end of the list is assumed as appropriate. For example:

```
>>> l = ["a", "b", "c", "d"]
>>> l[:3]
['a', 'b', 'c']
>>> l[3:]
['d']
>>>
```

You can also use negative indices to count back from the end of the list. The following example returns the last two elements in the list:

```
>>> l[-2:]
['c', 'd']
```

Incidentally, l[:-2] returns ['a', 'b'] in the preceding example.

See Also

All the recipes between Recipes 6.1 and 6.11 involve the use of lists.

See Recipe 5.15 where the same syntax is used for strings.

6.11. Applying a Function to a List

Problem

You need to apply a function to each element of a list and collect the results.

Solution

Use the Python language feature called *comprehensions*.

The following example will convert each string element of the list to uppercase and return a new list the same length as the old one, but with all the strings in uppercase:

```
>>> l = ["abc", "def", "ghi", "ijk"]
>>> [x.upper() for x in l]
['ABC', 'DEF', 'GHI', 'IJK']
```

Although it can get confusing, there is no reason why you can't combine these kinds of statements, nesting one comprehension inside another.

Discussion

This is a very concise way of doing comprehensions. The whole expression has to be enclosed in brackets ([]). The first element of the comprehension is the code to be evaluated for each element of the list. The rest of the comprehension looks rather like a normal list iteration command (Recipe 6.7). The loop variable follows the for keyword and then after the in keyword is the list to be used.

See Also

All the recipes between Recipes 6.1 and 6.11 involve the use of lists.

6.12. Creating a Dictionary

Problem

You need to create a lookup table where you associate values with keys.

Solution

Use a Python dictionary.

Arrays are great when you need to access a list of items in order, or you always know the index of the element that you want to use. Dictionaries are an alternative to lists for storing collections of data, but they are organized very differently.

Figure 6-1 shows how a dictionary is organized.

phone_numbers	
Key: Simon	Value: 01234 567899
Key: Jane	Value: 01234 666666
Key: Pete	Value: 01234 777555
Key: Linda	Value: 01234 887788

Figure 6-1. A Python dictionary

A dictionary stores key/value pairs in such a way that you can use the key to retrieve that value very efficiently and without having to search the whole dictionary.

To create a dictionary, you use the {} notation:

```
>>> phone_numbers = {'Simon':'01234 567899', 'Jane':'01234 666666'}
```

Discussion

In this example, the keys of the dictionary are strings, but they do not have to be; they could be numbers or in fact any data type, although strings are most commonly used.

The values can also be of any data type, including other dictionaries or lists. The following example creates one dictionary (a) and then uses it as a value in a second dictionary (b):

```
>>> a = {'key1':'value1', 'key2':2}
>>> a
{'key2': 2, 'key1': 'value1'}
>>> b = {'b_key1':a}
>>> b
{'b_key1': {'key2': 2, 'key1': 'value1'}}
```

When you display the contents of a dictionary, you will notice that the order of the items in the dictionary may not match the order in which they were specified when the dictionary was created and initialized with some contents:

```
>>> phone_numbers = {'Simon':'01234 567899', 'Jane':'01234 666666'}
>>> phone_numbers
{'Jane': '01234 666666', 'Simon': '01234 567899'}
```

Unlike lists, dictionaries have no concept of keeping items in order. Because of the way they are represented internally, the order of a dictionary's contents will be—for all intents and purposes—random.

The reason the order appears to be random is that the underlying data structure that is a *hash table*. Hash tables use a *hashing function* to decide where to store the value; the hashing function calculates a numeric equivalent to any object.

You can find out more about hash tables at Wikipedia (*http://en.wikipedia.org/wiki/Hash_table*).

See Also

All the recipes between Recipes 6.12 and 6.15 involve the use of dictionaries.

6.13. Accessing a Dictionary

Problem

You need to find and change entries in a dictionary.

Solution

Use the Python [] notation. Use the key of the entry to which you need access inside the brackets:

```
>>> phone_numbers = {'Simon':'01234 567899', 'Jane':'01234 666666'}
>>> phone_numbers['Simon']
'01234 567899'
>>> phone_numbers['Jane']
'01234 666666'
```

Discussion

If you use a key that is not present in the dictionary, you will get a *key error*. For example:

```
{'b_key1': {'key2': 2, 'key1': 'value1'}}
>>> phone_numbers = {'Simon':'01234 567899', 'Jane':'01234 666666'}
>>> phone_numbers['Phil']
Traceback (most recent call last):
  File "<stdin>", line 1, in <module>
```

```
KeyError: 'Phil'
>>>
```

As well as using the [] notation to read values from a dictionary, you can also use it to add new values or overwrite existing ones.

The following example adds a new entry to the dictionary with a key of `Pete` and a value of `01234 777555`:

```
>>> phone_numbers['Pete'] = '01234 777555'
>>> phone_numbers['Pete']
'01234 777555'
```

If the key is not in use in the dictionary, a new entry is automatically added. If the key is already present, then whatever value was there before will be overwritten by the new value.

This is in contrast to trying to read a value from the dictionary, where an unknown key will cause an error.

See Also

All the recipes between Recipes 6.12 and 6.15 involve the use of dictionaries.

For information on handling errors, see Recipe 7.10.

6.14. Removing Things from a Dictionary

Problem

You need to remove an item from a dictionary.

Solution

Use the pop command, specifying the key for the item that you want to remove:

```
>>> phone_numbers = {'Simon':'01234 567899', 'Jane':'01234 666666'}
>>> phone_numbers.pop('Jane')
'01234 666666'
>>> phone_numbers
{'Simon': '01234 567899'}
```

Discussion

The pop command returns the value of the item removed from the dictionary.

See Also

All the recipes between Recipes 6.12 and 6.15 involve the use of dictionaries.

6.15. Iterating over Dictionaries

Problem

You need to do something to each of the items in the dictionary in turn.

Solution

Use the for command to iterate over the keys of the dictionary:

```
>>> phone_numbers = {'Simon':'01234 567899', 'Jane':'01234 666666'}
>>> for name in phone_numbers:
...     print(name)
...
Jane
Simon
```

Discussion

There are a couple of other techniques that you can use to iterate over a dictionary. The following form can be useful if you need access to the values as well as the keys:

```
>>> phone_numbers = {'Simon':'01234 567899', 'Jane':'01234 666666'}
>>> for name, num in phone_numbers.items():
...     print(name + " " + num)
...
Jane 01234 666666
Simon 01234 567899
```

See Also

All the recipes between Recipes 6.12 and 6.15 involve the use of dictionaries.

See the for command used elsewhere in Recipes 5.3, 5.21, 6.7, and 6.11.

Advanced Python

7.0. Introduction

In this chapter, we will explore some of the more advanced concepts in the Python language—in particular, object-oriented Python, reading and writing files, handling exceptions, using modules, and Internet programming.

7.1. Formatting Numbers

Problem

You want to format numbers to a certain number of decimal places.

Solution

Apply a `format` string to the number. For example:

```
>>> x = 1.2345678
>>> "x={:.2f}".format(x)
'x=1.23'
>>>
```

Discussion

The formatting string can contain a mixture of regular text and markers delimited by { and }. The parameters to the format function (there can be as many as you like) will be substituted in place of the marker, according to the format specifier.

In the preceding example, the format specifier is `:.2f`, which means that the number will be specified with two digits after the decimal place and is a float `f`.

If you wanted the number to be formatted so the total length of the number is always seven digits (or padding spaces), then you would add another number before the decimal place like this:

```
>>> "x={:7.2f}".format(x)
'x=   1.23'
>>>
```

In this case, since the number is only three digits long, there are four spaces of padding before the 1.

A more complicated example might be to display the temperature in both degrees Celsius and degrees Fahrenheit, as shown here:

```
>>> c = 20.5
>>> "Temperature {:5.2f} deg C, {:5.2f} deg F.".format(c, c * 9 / 5 + 32)
'Temperature 20.50 deg C, 68.90 deg F.'
>>>
```

See Also

Formatting in Python involves a whole formatting language (*http://bit.ly/1cwzPp5*).

7.2. Formatting Dates

Problem

You want to convert a date into a string and format it in a certain way.

Solution

Apply `format` string to the date object.

For example:

```
>>> from datetime import datetime
>>> d = datetime.now()
>>> "{:%Y-%m-%d %H:%M:%S}".format(d)
'2013-05-02 16:00:45'
>>>
```

Discussion

The Python formatting language includes some special symbols for formatting date. `%Y`, `%m`, and `%d` correspond to year, month, and day numbers, respectively.

See Also

See Recipe 7.1 for formatting of numbers.

Formatting in Python actually involves a whole formatting language. You will find full details of this at Python's website (*http://bit.ly/1cwzPp5*).

7.3. Returning More Than One Value

Problem

You need to write a function that returns more than one value.

Solution

Return a Python *tuple* and use the multiple variable assignment syntax.

For example:

```
>>> def calculate_temperatures(kelvin):
...     celsius = kelvin - 273
...     fahrenheit = celsius * 9 / 5 + 32
...     return celsius, fahrenheit
...
>>> c, f = calculate_temperatures(340)
>>>
>>> print(c)
67
>>> print(f)
152.6
```

A tuple is a Python data structure that's a little like a list, except that tuples are enclosed in parentheses rather than brackets. They are also of fixed size.

Discussion

Sometimes, when you just have one or two values to return, this is the right way to return multiple values. However, if the data is complex, you might find the solution is neater if you use Python's object-oriented features and define a class that contains the data. That way, you can return an instance of the class rather than a tuple.

See Also

See Recipe 7.4 for information on defining classes.

7.4. Defining a Class

Problem

You need to group together related data and functionality into a class.

Solution

Define a class and provide it with the member variables you need.

The following example defines a class to represent an address book entry:

```
class Person:
    '''This class represents a person object'''

    def __init__(self, name, tel):
        self.name = name
        self.tel = tel
```

The first line inside the class definition uses the triple, single, or double quotes to denote a documentation string. This should explain the purpose of the class. Although entirely optional, adding a documentation string to a class allows others to see what the class does. This is particularly useful if the class is made available for others to use.

Doc strings are not like normal comments because, although they are not active lines of code, they do get associated with the class, so, at any time, you can read the doc string for a class using the following command:

```
Person.__doc__
```

Inside the class definition is the constructor method, which will be called automatically whenever you create a new instance of the class. A class is like a template, so in defining a class called `Person`, we do not create any actual `Person` objects until later:

```
def __init__(self, name, tel):
    self.name = name
    self.tel = tel
```

The constructor method must be named as shown with double underscores on either side of the word `init`.

Discussion

One way in which Python differs from most object-oriented languages is that you have to include the special variable `self` as a parameter to all the methods that you define within the class. This is a reference to, in this case, the newly created instance. The variable `self` is the same concept as the special variable `this` that you find in Java and some other languages.

The code in this method transfers parameters that were supplied to it into member variables. The member variables do not need to be declared in advance, but do need to be prefixed by `self.`.

So the following line:

```
self.name = name
```

creates a variable called name that's accessible to every member of the class `Person` and initializes it with the value passed into the call to create an instance, which looks like this:

```
p = Person("Simon", "1234567")
```

We can then check that our new `Person` object, p, has a name of "Simon" by typing the following:

```
>>> p.name
Simon
```

In a complex program, it is good practice to put each class in its own file with a filename that matches the class name. This also makes it easy to convert the class into a module (see Recipe 7.11).

See Also

See Recipe 7.5 for information on defining methods.

7.5. Defining a Method

Problem

You need to add a method to a class.

Solution

The following example shows how you can include a method within a class definition:

```
class Person:
    '''This class represents a person object'''

    def __init__(self, first_name, surname, tel):
        self.first_name = first_name
        self.surname = surname
        self.tel = tel

    def full_name(self):
        return self.first_name + " " + self.surname
```

The `full_name` method concatenates the first name and surname attributes of the person, with a space in between.

Discussion

You can think of methods as just functions that are tied to a specific class and may or may not use member variables of that class in their processing. So, as with a function, you can write whatever code you like in a method and also have one method call another.

If one method calls another within the same class, the call to the method has to be prefixed with self..

See Also

See Recipe 7.4 for information on defining a class.

7.6. Inheritance

Problem

You need a specialized version of an existing class.

Solution

Use *inheritance* to create a subclass of an existing class and add new member variables and methods.

By default, all new classes that you create are subclasses of object. You can change this by specifying the class you want to use as a superclass in parentheses after the class name in a class definition. The following example defines a class (Employee) as a subclass (Person) and adds a new member variable (salary) and an extra method (give_raise):

```
class Employee(Person):

    def __init__(self, first_name, surname, tel, salary):
        super().__init__(first_name, surname, tel)
        self.salary = salary

    def give_raise(self, amount):
        self.salary = self.salary + amount
```

Note that the preceding example is for Python 3. For Python 2, you can't use super the same way. Instead, you must write:

```
class Employee(Person):

    def __init__(self, first_name, surname, tel, salary):
        Person.__init__(self, first_name, surname, tel)
        self.salary = salary

    def give_raise(self, amount):
        self.salary = self.salary + amount
```

Discussion

In both these examples, the initializer method for the subclass first uses the initializer method of the parent class (superclass) and then adds the member variable. This has the advantage that you do not need to repeat the initialization code in the new subclass.

See Also

See Recipe 5.24 for information on defining a class.

The Python inheritance mechanism is actually very powerful and supports *multiple inheritance*—where a subclass inherits from more than one superclass. For more on inheritance, see the official documentation for Python (*http://bit.ly/17Y2c2k*).

7.7. Writing to a File

Problem

You need to write something to a file.

Solution

Use the open, write, and close functions to open a file, write some data, and then close the file, respectively:

```
>>> f = open('test.txt', 'w')
>>> f.write('This file is not empty')
>>> f.close()
```

Discussion

Once you have opened the file, you can make as many writes to it as you like before you close the file. Note that it is important to use close because although each write should update the file immediately, it may be buffered in memory and data could be lost. It could also leave the file locked so that other programs can't open it.

The open method takes two parameters. The first is the path to the file to be written. This can be relative to the current working directory, or if it starts with a /, an absolute path.

The second parameter is the mode in which the file should be opened. To overwrite an existing file, or create the file with the name specified if it doesn't exist, just use w. Table 7-1 shows the full list of file mode characters. You can combine these using +. So to open a file in read and binary mode, you would use:

```
>>> f = open('test.txt', 'r+b')
```

Table 7-1. File modes

Mode	Description
r	Read
w	Write
a	Append - to the end of an existing file rather than overwriting
b	Binary mode
t	Text mode (default)
+	A shortcut for r+w

Binary mode allows you to read or write binary streams of data, such as images, rather than text.

See Also

To read the contents of a file, see Recipe 7.8. For more information on handling exceptions, see Recipe 7.11.

7.8. Reading from a File

Problem

You need to read the contents of a file into a string variable.

Solution

To read a file's contents, you need to use the file methods `open`, `read`, and `close`. The following example reads the entire contents of the file into the variable `s`:

```
f = open('test.txt')
s = f.read()
f.close()
```

Discussion

You can also read text files one line at a time using the method `readline`.

The preceding example will throw an exception if the file does not exist or is not readable for some other reason. You can handle this by enclosing the code in a `try/except` construction like this:

```
try:
    f = open('test.txt')
    s = f.read()
    f.close()
```

```
    except IOError:
        print("Cannot open the file")
```

See Also

To write things to a file and for a list of file open modes, see Recipe 7.7.

For more information on handling exceptions, see Recipe 7.10.

7.9. Pickling

Problem

You want to save the entire contents of a data structure to a file so that it can be read the next time the program is run.

Solution

Use the Python *pickling* feature to dump the data structure to file in a format that can be automatically read back into memory as an equivalent data structure later on.

The following example saves a complex list structure to file:

```
>>> import pickle
>>> mylist = ['some text', 123, [4, 5, True]]
>>> f = open('mylist.pickle', 'w')
>>> pickle.dump(mylist, f)
>>> f.close()
```

To *unpickle* the contents of the file into a new list, use the following:

```
>>> f = open('mylist.pickle')
>>> other_array = pickle.load(f)
>>> f.close()
>>> other_array
['some text', 123, [4, 5, True]]
```

Discussion

Pickling will work on pretty much any data structure you can throw at it. It does not need to be a list.

The file is saved in a text format that is sort of human-readable, but you will not normally need to look at or edit the text file.

See Also

To write things to a file and for a list of file open modes, see Recipe 7.6.

7.10. Handling Exceptions

Problem

If something goes wrong while a program is running, you want to catch the error or exception and display a more user-friendly error message.

Solution

Use Python's `try`/`except` construct.

The following example, taken from Recipe 7.7, catches any problems in opening a file:

```
try:
    f = open('test.txt')
    s = f.read()
    f.close()
except IOError:
    print("Cannot open the file")
```

Discussion

A common situation where runtime exceptions can occur—in addition to during file access—is when you are accessing a list and the index you are using is outside the bounds of the list. For example, this happens if you try to access the fourth element of a three-element list:

```
>>> list = [1, 2, 3]
>>> list[4]
Traceback (most recent call last):
  File "<stdin>", line 1, in <module>
IndexError: list index out of range
```

Errors and exceptions are arranged in a hierarchy, and you can be as specific or general as you like when catching the exceptions.

`Exception` is pretty near the top of that tree and will catch almost any exception. You can also have separate `except` sections for catching different types of exception and handling each in a different way. If you do not specify any exception class, all exceptions will be caught.

Python also allows you to have `else` and `finally` clauses in your error handling:

```
list = [1, 2, 3]
try:
    list[8]
except:
    print("out of range")
else:
    print("in range")
```

```
finally:
    print("always do this")
```

The `else` clause will be run if there is no exception, and the `finally` clause will be run whether there is an exception or not.

Whenever an exception occurs, you can get more information about it using the exception object, which is available only if you use the `as` keyword, as shown in the following example:

```
>>> list = [1, 2, 3]
>>> try:
...     list[8]
... except Exception as e:
...     print("out of range")
...     print(e)
...
out of range
list index out of range
>>>
```

See Also

Python documentation (*http://bit.ly/1cqqHV7*) for Python exception class hierarchy.

7.11. Using Modules

Problem

You want to use a Python module in your program.

Solution

Use the `import` command:

```
import random
```

Discussion

There is a very large number of modules (sometimes called *libraries*) available for Python. Many are included with Python as standard and others can be downloaded and installed into Python.

Standard Python libraries include modules for random numbers, database access, various Internet protocols, object serialization, and many more.

One consequence of having so many modules is that there is the potential for conflict—for example, if two modules have a function of the same name. To avoid such conflicts, when importing a module, you can specify how much of the module is accessible.

For example, if you just use a command like this:

```
import random
```

there is no possibility of any conflicts because you will only be able to access any functions or variables in the module by prefixing them with random.. (Incidentally, we will be meeting the random package in the next recipe.)

If, on the other hand, you use the following command, everything in the module will be accessible without your having to put anything in front of it. Unless you know what all the functions are in all the modules you are using, there is a much greater chance of conflict.

```
from random import *
```

In between these two extremes, you can explicitly specify those components of a module that you need within a program so that they can be conveniently used without any prefix.

For example:

```
>>> from random import randint
>>> print(randint(1,6))
2
>>>
```

A third option is to use the as keyword to provide a more convenient or meaningful name for the module when referencing it.

```
>>> import random as R
>>> R.randint(1, 6)
```

See Also

For a definitive list of the modules included with Python, see *http://docs.python.org/2/library/*.

7.12. Random Numbers

Problem

You need to generate a random number between a range of numbers.

Solution

Use the random library:

```
>>> import random
>>> random.randint(1, 6)
2
>>> random.randint(1, 6)
```

```
6
>>> random.randint(1, 6)
5
```

The number generated will be between the two arguments (inclusive)--in this case, simulating a gaming dice.

Discussion

The numbers generated are not truly random, but are what is known as a *pseudorandom number sequence*; that is, they are a long sequence of numbers that, when taken in a large enough quantity, show what statisticians call a *random distribution*. For games, this is perfectly good enough, but if you were generating lottery numbers, you would need to look at special randomizing hardware. Computers are just plain bad at being random; it's not really in their nature.

A common use of random numbers is to select something at random from a list. You can do this by generating an index position and using that, but there is also a command in the random module specifically for this. Try out the following example:

```
>>> import random
>>> random.choice(['a', 'b', 'c'])
'a'
>>> random.choice(['a', 'b', 'c'])
'b'
>>> random.choice(['a', 'b', 'c'])
'a'
```

See Also

The random library has some other useful functions for doing things like picking from lists at random. See the official reference at *http://docs.python.org/2/library/ random.html* for more information on this.

7.13. Making Web Requests from Python

Problem

You need to read the contents of a web page into a string using Python.

Solution

Python has an extensive library for making HTTP requests.

The following example reads the contents of the Google home page into the string contents:

```
import urllib2
contents = urllib2.urlopen("https://www.google.com/").read()
print(contents)
```

Discussion

Having read the HTML, you are then likely to want to search it and extract the parts of the text that you really want. For this, you will need to use the string manipulation functions (see Recipes 5.14 and 5.15).

See Also

For an example of using web requests to check for Gmail messages, see Recipe 7.15.

7.14. Command-Line Arguments in Python

Problem

You want to run a Python program from the command line and pass it parameters.

Solution

Import sys and use its argv property, as shown in the following example. This returns an array, the first element of which is the name of the program. The other elements are any parameters (separated by spaces) that were typed on the command line after the program name.

```
import sys

for (i, value) in enumerate(sys.argv):
    print("arg: %d %s " % (i, value))
```

Running the program from the command line, with some parameters after it, results in the following output:

```
$ python cmd_line.py a b c
arg: 0 cmd_line.py
arg: 1 a
arg: 2 b
arg: 3 c
```

Discussion

Being able to specify command-line arguments can be useful for automating the running of Python programs, either at startup (Recipe 3.20) or on a timed basis (Recipe 3.21).

See Also

For basic information on running Python from the command line, see Recipe 5.4.

In printing out `argv`, we used list enumerations (Recipe 6.8).

7.15. Sending Email from Python

Problem

You want to send an email message from a Python program.

Solution

Python has a library for the Simple Mail Transfer Protocol (SMTP) that you can use to send emails:

```python
import smtplib

GMAIL_USER = 'your_name@gmail.com'
GMAIL_PASS = 'your_password'
SMTP_SERVER = 'smtp.gmail.com'
SMTP_PORT = 587

def send_email(recipient, subject, text):
    smtpserver = smtplib.SMTP(SMTP_SERVER, SMTP_PORT)
    smtpserver.ehlo()
    smtpserver.starttls()
    smtpserver.ehlo
    smtpserver.login(GMAIL_USER, GMAIL_PASS)
    header = 'To:' + recipient + '\n' + 'From: ' + GMAIL_USER
    header = header + '\n' + 'Subject:' + subject + '\n'
    msg = header + '\n' + text + ' \n\n'
    smtpserver.sendmail(GMAIL_USER, recipient, msg)
    smtpserver.close()

send_email('destination_email_address', 'sub', 'this is text')
```

To use this example to send an email to an address of your choice, first change the variables `GMAIL_USER` and `GMAIL_PASS` to match your email credentials. If you are not using Gmail, then you will also need to change the values of the `SMTP_SERVER` and possibly `SMTP_PORT`.

You will also need to change the destination email address in the last line.

Discussion

The `send_email` message simplifies the use of the `smtplib` library into a single function that you can reuse in your projects.

Being able to send emails from Python opens up all sorts of project opportunities. For example, you could use a sensor such as the PIR sensor of Recipe 11.9 to send an email when movement is detected.

See Also

For performing HTTP requests from the Raspberry Pi, see Recipe 7.13.

For more information on the `smtplib`, see *http://docs.python.org/2/library/smtplib.html*.

7.16. Writing a Simple Web Server in Python

Problem

You need to create a simple Python web server, but don't want to have to run a full web server stack.

Solution

Use the `bottle` Python library to run a pure Python web server that will respond to HTTP requests.

To install `bottle`, use the following command:

```
sudo apt-get install python-bottle
```

The following Python program (called bottle_test in the book resources (*http://www.raspberrypicookbook.com*)) simply serves up a message displaying what time the Pi thinks it is. Figure 7-1 shows the page you see if you connect to the Raspberry Pi from a browser anywhere on your network:

```
from bottle import route, run, template
from datetime import datetime

@route('/')
def index(name='time'):
    dt = datetime.now()
    time = "{:%Y-%m-%d %H:%M:%S}".format(dt)
    return template('<b>Pi thinks the date/time is: {{t}}</b>', t=time)

run(host='192.168.1.16', port=80)
```

To start the program, you need to run it with superuser privileges:

```
$ sudo python bottle_test.py
```

This example requires a little explanation.

Figure 7-1. Browsing to a Python bottle web server

After the `import` commands, the `@route` command links the URL path / with the handler function that follows it.

That handler function formats the date and time and then returns a string of the HTML to be rendered by the browser. In this case, it uses a template into which values can be substituted.

The final `run` line actually starts the web serving process. Note that you need to specify the hostname and port. Port 80 is the default port for web serving, so if you wish to use a different port, then you need to add the port number with a : after the server address.

Discussion

You can define as many routes and handlers as you like within the program.

`bottle` is perfect for a small, simple web server projects, and because it's written in Python, it's very easy to write a handler function to control hardware in response to the user interacting with the page in a browser.

See Also

For more information, see the `bottle` project documentation (*http://bottlepy.org/docs/ dev/*).

For more on formatting dates and times in Python, see Recipe 7.2.

GPIO Basics

8.0. Introduction

This chapter contains some basic recipes for setting up and using the Raspberry Pi's general-purpose input/output (GPIO) connector.

8.1. Finding Your Way Around the GPIO Connector

 Be sure to check out the accompanying video for this recipe at *http://razzpisampler.oreilly.com*.

Problem

You need to connect electronics to the GPIO connector, but you need to know more about what all the pins do.

Solution

Figure 8-1 shows the GPIO pinout for both revisions 1 and 2 of the Raspberry Pi Model B. The quick way to tell the boards apart is that if you have an older revision 1 board, it has a black audio socket. The revision 2 boards have a blue audio socket.

Figure 8-1. The GPIO pinout

There were three changes to the GPIO connector between revision 1 and revision 2. These are highlighted in bold in Figure 8-1. First, the I2C port was swapped. The two pins SDA and SCL are still SDA and SCL but use a different internal I2C interface. This means that if you're using the pins as GPIO rather than I2C, then you will refer to them as 2 and 3 on a revision 2 board. Also, GPIO 21 was replaced by GPIO 27 on revision 2.

At the top of the connector, there are 3.3V and 5V power supplies. The GPIO uses 3.3V for all inputs and outputs. Any pin with a number next to it can act as a GPIO pin. Those that have another name after them also have some other special purpose, so 14 TXD and 15 RXD are the transmit and receive pins of the serial interface. SDA and SCL form the I2C interface, and MOSI, MISO, and SCKL from the SPI interface.

Discussion

A GPIO pin can be used as either a digital input or a digital output, and both operate at 3.3V. Unlike the Arduino, the Raspberry Pi does not have any analog inputs. For that you must use an external analog-to-digital converter (ADC) or connect the Pi to an interface board (like the Gertboard) or to an Arduino or an aLaMode board, as discussed in Chapter 14.

See Also

See Recipe 12.4 for connecting an ADC to a Raspberry Pi.

8.2. Keeping Your Raspberry Pi Safe when Using the GPIO Connector

Problem

You want to connect external electronics to your Raspberry Pi, but don't want to accidentally damage or break it.

Solution

Obey these simple rules to reduce the risk of damaging your Raspberry Pi when using the GPIO connector:

- Do not put more than 3.3V on any GPIO pin.
- Do not draw more than 3mA per output. You can *get away* with drawing more, but this may shorten the life of the Pi. 3mA is enough to light a red LED with a 470Ω series resistor.
- Do not poke at the GPIO connector with a screwdriver or any metal object when the Pi is powered up.
- Do not power the Pi with more than 5V.
- Do not draw more than a total of 50mA from the 3.3V supply pins.
- Do not draw more than a total of 250mA from the 5V supply pins.

Discussion

There is no doubt about it, the Raspberry Pi is a little fragile when it comes to adding external electronics. Exercise caution and check what you have done *before* you power up the Raspberry Pi, or run the risk of having to replace your Raspberry Pi.

See Also

Read this very good discussion of the Raspberry Pi's GPIO output capabilities (*http://bit.ly/1aMQ0fh*).

8.3. Installing RPi.GPIO

Problem

You want to be able to set the output of GPIO pins and read the input values using Python.

Solution

Download and install the RPi.GPIO Python library.

From a Terminal window on your Pi, type the following commands to fetch and install the RPi.GPIO library.

```
$ sudo apt-get install python-dev
$ sudo apt-get install python-rpi.gpio
```

Note that the latest versions of many distributions may have RPi.GPIO already installed. If this is the case, then the preceding commands will just update it to the latest version.

Discussion

This library will be used throughout the book. The library uses native C wrapped in Python to make the IO as fast as possible. However, the Pi is not really designed as a microcontroller, so the GPIO pins will not respond as quickly as an Arduino.

See Also

An alternative library that performs much the same job as RPi.GPIO is WiringPi. You can find out more about the library at Gordons Projects (*http://bit.ly/1bU7aKu*).

Many of the recipes in this book use the RPi.GPIO library.

8.4. Setting Up I2C

Problem

You have an I2C device that you want to use with your Raspberry Pi, so you want to know how to make I2C work with the Raspberry Pi.

Solution

If you are using Adafruit Occidentalis 0.2 or later, then there is nothing to do. The distribution is preconfigured with I2C support.

Newer versions of the distributions may make these steps unneccessary. But in any case, this will ensure that you get the latest version.

If you are using Raspbian, there are a couple of configuration changes you need to make.

Edit the file */etc/modules* using the commmand sudo nano /etc/modules and add the following lines to the end of it:

```
i2c-bcm2708
i2c-dev
```

You may also need to edit the file */etc/modprobe.d/raspi-blacklist.conf* and comment out this line:

```
blacklist i2c-bcm2708
```

by adding a # character in front of it:

```
#blacklist i2c-bcm2708
```

If you plan to use SPI (Recipe 8.6), you may want to comment out the SPI line from the *blacklist* as well.

To install the Python I2C library, run the following command:

```
$ sudo apt-get install python-smbus
```

Reboot your Pi and it will be ready for I2C.

Discussion

Using I2C modules is actually a really good way of interfacing with the Pi. It reduces the number of wires that you need to connect everything up (to just four), and there are some really neat I2C modules available.

However, don't forget to total up the current used by the I2C modules and make sure that it doesn't exceed those specified in Recipe 8.2.

Figure 8-2 shows a selection of I2C modules available from Adafruit. Other suppliers, such as SparkFun, also have I2C devices. From left to right, there are LED matrix displays; a four-digit, seven-segment LED display; a 16-channel PWM/servo controller; and a real-time clock module.

Figure 8-2. I2C modules

Other I2C modules available include FM radio transmitters, ultrasonic rangefinders, OLED displays, and various types of sensors.

See Also

See some of the I2C recipes in this book: Recipes 10.2, 13.1, 13.2, and 13.4.

8.5. Using I2C Tools

Problem

You have an I2C device attached to your Raspberry Pi, and you want to know how to check that it's attached and to find its I2C address.

Solution

Install and use `i2c-tools`.

 On newer distributions, you may find that `i2c-tools` is already installed.

From a Terminal window on your Pi, type the following commands to fetch and install the `i2c-tools`:

```
$ sudo apt-get install i2c-tools
```

Attach your I2C device to the Pi and run the command:

```
$ sudo i2cdetect -y 1
```

Note that if you are using an older revision 1 board, you need to change 1 to 0 in the preceding code line.

If I2C is available, you will see some output like that shown in Figure 8-3. This shows that two I2C addresses are in use—0x40 and 0x70.

Figure 8-3. i2C-tools

Discussion

i2cDetect is a useful diagnostic tool and one that is worth running the first time you use a new I2C device.

See Also

See some of the I2C recipes in this book: Recipes 10.2, 13.1, 13.2, and 13.4.

For more information on installing with `apt-get`, see Recipe 3.16.

8.6. Setting Up SPI

Problem

You have an SPI (serial peripheral interface) bus that you want to use with your Raspberry Pi.

Solution

If you are using Adafruit Occidentalis 0.2 or later, there is nothing to do. The distribution is preconfigured with SPI support.

If you are using Raspbian, there are a couple of configuration changes you need to make.

Edit the file */etc/modules* using the command `sudo nano /etc/modules` and add the following line to the end of it:

```
spidev
```

You may also need to edit the file *etc/modprobe.d/raspi-blacklist.conf* and comment out the line:

```
blacklist spi-bcm2708
```

by adding a # character in front of it:

```
#blacklist spi-bcm2708
```

If you plan to use I2C, you may want to comment out the I2C line from the *blacklist* as well.

The SPI feature of the Raspberry Pi is also supported by a Python library that will allow you to carry out SPI communication from a Python program. To install this, first install Git (Recipe 3.19), and then issue the following commands:

```
$ cd ~
$ sudo apt-get install python-dev
$ git clone git://github.com/doceme/py-spidev
$ cd py-spidev/
$ sudo python setup.py install
```

Reboot your Pi and it will be ready for SPI.

Discussion

SPI allows serial transfer of data between the Raspberry Pi and peripheral devices, such as analog-to-digital converter (ADC) and port expander chips, among other devices.

You may come across some examples of interfacing to SPI that do not use the SPI interface but instead use an approach called *bit banging*, where the RPi.GPIO library is used to interface with the four GPIO pins used by the SPI interface.

See Also

We use a SPI analog-to-digital converter chip in Recipe 12.4.

8.7. Freeing the Serial Port

Problem

You want to use the serial port (Rx and Tx pins) on the Raspberry Pi for your own projects, but it is in use by Linux as a console connection.

Solution

By default, the serial port acts as a console, through which you can connect to the Raspberry Pi using a special serial lead (see Recipe 2.6).

To disable this so that you can use the serial port to connect to peripherals such as GPS (Recipe 11.10), comment out a line in *etc/inittab*:

```
$ sudo nano /etc/inittab
```

Scroll down to the end of the file to find the line:

```
T0:23:respawn:/sbin/getty -L ttyAMA0 115200 vt100
```

Comment it out by placing a # in front of it:

```
#T0:23:respawn:/sbin/getty -L ttyAMA0 115200 vt100
```

Save the file with Ctrl-X followed by Y.

For the changes to take effect, you need to reboot your Raspberry Pi with `sudo reboot`.

Discussion

One effect of making this change is that you will no longer be able to connect to the Raspberry Pi with a serial console cable. However, it will not affect WiFi or Ethernet connections.

If, in the future, you want to reverse this operation, then just change the files back to how they were and reboot.

See Also

You will need to use this technique in recipes that connect hardware to the serial port, such as Recipe 11.10 and many of the recipes in Chapter 14 that communicate with an Arduino using the serial port.

8.8. Installing PySerial for Access to the Serial Port from Python

Problem

You want to use the serial port (Rx and Tx pins) on the Raspberry Pi using Python.

Solution

Install the PySerial library:

```
$ sudo apt-get install python-serial
```

Before you can use the library for your own Python serial projects, you need to disable the Raspberry Pi's serial console by following Recipe 8.7.

Discussion

The library is pretty easy to use. Creating a connection uses the following syntax:

```
ser = serial.Serial(DEVICE, BAUD)
```

Where **DEVICE** is the device for the serial port (*/dev/ttyAMA0*) and **BAUD** the baud rate as a number, not a string. For example:

```
ser = serial.Serial('/dev/ttyAMA0', 9600)
```

Once a connection is established, you can send data over serial like this:

```
ser.write('some text')
```

Listening for a response normally involves a loop that reads and prints, as illustrated in this example:

```
while True:
    print(ser.read())
```

See Also

You will need to use this technique in recipes that connect hardware to the serial port, such as Recipe 11.10.

8.9. Installing Minicom to Test the Serial Port

Problem

You want to send and receive serial commands from a Terminal session.

Solution

Install Minicom:

```
$ sudo apt-get install minicom
```

Before you can use Minicom for serial communication on your own projects, you need to disable the Raspberry Pi's serial console by following Recipe 8.7.

Once Minicom is installed, you can start a serial communications session with a serial device connected to the RXD and TXD pins of the GPIO connector using this command:

```
$ minicom -b 9600 -o -D /dev/ttyAMA0
```

The parameter after -b is the baud rate, and after -D the serial port. Remember to use the same baud rate as set on the device you are communicating with.

This will start a Minicom session. One of the first things you want to do is turn on *local echo* so you can see the command that you are typing. To do this, press Ctrl-A and then Z; you'll see the menu shown in Figure 8-4. Press E to turn on local echo.

```
  ⌂ si — pi@raspberrypi: ~ — ssh — 81×24

Welco+--------------------------------------------------------+
      |                  Minicom Command Summary               |
OPTIO |                                                        |
Compi |          Commands can be called by CTRL-A <key>        |
Port  |                                                        |
      |              Main Functions          Other Functions   |
Press |                                                        |
      |    Dialing directory..D  run script (Go)....G | Clear Screen.......C |
      |    Send files.........S  Receive files......R | cOnfigure Minicom..O |
      |    comm Parameters....P  Add linefeed.......A | Suspend minicom....J |
      |    Capture on/off.....L  Hangup............H | eXit and reset.....X |
      |    send break........F   initialize Modem...M | Quit with no reset.Q |
      |    Terminal settings..T  run Kermit.........K | Cursor key mode....I |
      |    lineWrap on/off....W  local Echo on/off..E | Help screen........Z |
      |    Paste file........Y   Timestamp toggle...N | scroll Back........B |
      |                                                        |
      |        Select function or press Enter for none.        |
      |                                                        |
      |        Written by Miquel van Smoorenburg 1991-1995     |
      |        Some additions by Jukka Lahtinen 1997-2000      |
      |        i18n by Arnaldo Carvalho de Melo 1998           |
      +--------------------------------------------------------+
 CTRL-A Z for help  |  9600 8N1  | NOR | Minicom 2.6.1  | VT102 |    Offline
```

Figure 8-4. Minicom commands

Now anything you type will be sent to the serial device, and all messages coming from the device will also be displayed.

Discussion

Minicom is a great tool for checking out the messages coming from a serial device or making sure that it's working. Having established that you will probably want to write a Python program to handle the serial communications, you will need the Python serial library (Recipe 8.8).

See Also

Check out the Minicom documentation (*http://linux.die.net/man/1/minicom*). For an example of using Minicom, see Recipe 13.3.

8.10. Using a Breadboard with Jumper Leads

Problem

You want to do some electronic prototyping using your Raspberry Pi and solderless breadboard.

Solution

Use male-to-female jumper wires and a paper pin label template (Figure 8-5).

Figure 8-5. Connecting Pi to breadboard using male-to-female jumper wires

Discussion

It's not always easy to identify the pins you want on a bare Raspberry Pi board. You can greatly simplify this by printing out a paper template to fit over the pins. One such template is available from Doctor Monk (*http://bit.ly/17gtRIV*).

You will also find it useful to have a selection of male-to-male jumper wires for making connections from one part of the breadboard to another.

Female-to-female jumper wires are also useful for connecting modules with male header pins directly to the Raspberry Pi, where there are no other components that might warrant the use of breadboard. This is actually the case for Figure 8-5.

See Also

This approach works well when you have a small number of connections. When you have more to make, then a Pi Cobbler is better (see Recipe 8.11).

8.11. Using a Breadboard with a Pi Cobbler

Problem

You want to do some electronic prototyping using a Raspberry Pi and solderless breadboard.

Solution

Use a Pi Cobbler, a device consisting of a small printed circuit board (PCB) with header pins that are designed to fit into a solderless breadboard like a dual inline package (DIL) chip. The pins are all labeled and the PCB also has a header socket. The supplied ribbon cable is used to link the Cobbler itself to the Raspberry Pi (see Figure 8-6).

Discussion

A great advantage of the Cobbler is that you can assemble the components onto the breadboard and then plug in the ribbon cable when it's ready.

Make sure that the red edge of the ribbon cable is toward the SD card edge of the Raspberry Pi. The cable only fits into the socket on the Pi Cobbler the correct way around.

See Also

See also Recipe 8.10.

8.12. Converting 5V Signals to 3.3V with Two Resistors

Problem

The Raspberry Pi operates at 3.3V, but you want to know how to connect the 5V output of a module to a GPIO pin on the Pi without damaging it.

Figure 8-6. Connecting Pi to breadboard using a Pi Cobbler

Solution

Use a pair of resistors as a potential divider to reduce the output voltage. Figure 8-7 shows how you might use the 5V serial connection of an Arduino to a Raspberry Pi.

To make this recipe, you will need:

- 270Ω resistor ("Resistors and Capacitors" on page 380)
- 470Ω resistor ("Resistors and Capacitors" on page 380)

The TXD signal from the Pi is a 3.3V output. This can be connected directly to a 5V input on the Arduino without any problem. The Arduino module will recognize anything over about 2.5V as being high.

The problem arises when you need to connect the 5V output of the Arduino module to the RXD pin of the Pi. This *must not* be connected directly to the RXD input. The 5V signal could damage the Pi. Instead, the two resistors shown in Figure 8-7 are used.

Discussion

The resistors used here will draw a current of 6mA. Given that the Pi uses a fairly hefty 500mA, this will not noticeably affect the current consumption of the Pi.

Figure 8-7. Using resistors to convert a 5V signal to 3.3V

If you want to minimize the current used by the potential divider, then use larger value resistors, scaled proportionally—for example, 27kΩ and 47kΩ, which will draw a miserly 60 μA.

See Also

For more information on connecting Raspberry Pi to Arduino, see Chapter 14.

If you have multiple signals to convert between 3.3V and 5V, then it's probably best to use a multichannel level converter module—see Recipe 8.13.

8.13. Converting 5V Signals to 3.3V with a Level Converter Module

Problem

The Raspberry Pi operates at 3.3V. You want to connect a number of 5V digital pins to a GPIO pin on the Pi without damaging it.

Solution

Use a bidirectional level converter module, such as one shown in Figure 8-8.

These modules are very easy to use. One side has the power supply at one voltage and a number of channels that can be either inputs or outputs at that voltage. The pins on the other side of the module have a power pin at the second voltage and all the inputs and outputs on that side are automatically converted to the voltage level for that side.

Figure 8-8. Level converter modules

Discussion

These level converters are available with differing numbers of channels. The two shown in Figure 8-8 have four and eight channels.

Sources for such level converters can be found in "Modules" on page 381.

See Also

See Recipe 8.12, especially if you have only one or two levels to convert.

8.14. Powering a Raspberry Pi with Batteries

Problem

You want to attach your Raspberry Pi to a robot and power it from alkaline batteries.

Solution

A typical project using a Raspberry Pi requires 5V at up to about 600mA (see Recipe 1.3). The requirement for 5V is strict. You should not try to power your Pi from more or less than 5V. In practice, this means you are most likely to power your Pi from a battery at a higher voltage than 5V—say, 9V—and use a voltage regulator to drop the voltage to 5V.

The relatively high power requirement of the Pi makes it unsuitable for powering from, say, a small 9V battery or even a set of AAA batteries. Most suitable would be a pack made up of six rechargeable AA batteries and a voltage regulator.

Figure 8-9 shows how you can use a voltage regulator with a battery pack to power a Pi through the 5V pin on the GPIO connector.

Figure 8-9. Powering a Raspberry Pi with AA batteries

Discussion

The 7805 voltage regulator will get pretty hot. If it gets too hot, its thermal cutout will kick in and the voltage at its output will drop, probably causing the Pi to reset.

A heat sink that clips onto the integrated circuit (IC) will help with this.

The 7805 requires the input voltage to be at least 2V above 5V. You can also buy low dropout (LDO) regulators such as the LM2940. The LM2940 has the same pinout as the 7805, but only requires the input to be 0.5V above the 5V output. However, remember that nominally, 1.5V AA cells soon drop to about 1.2V. So a pack of four of them is unlikely to provide enough voltage for more than a few minutes. A pack of six will be fine.

If you want to fit your Raspberry Pi into an automobile or RV, this recipe will work. You will also need to use a small monitor that can be powered from DC (direct current). Such devices are fairly easily found, as they are commonly used with closed-circuit camera systems.

See Also

You could also look out for a ready-made cellphone charger that accepts AA batteries and has a mini-USB socket on the end. Make sure that it can cope with 600mA or more.

Recipe 8.15 shows you how to power a Raspberry Pi from a LiPo battery pack.

8.15. Powering a Raspberry Pi with a LiPo Battery

Problem

You want to attach your Raspberry Pi to a robot and power it from a 3.7V LiPo battery.

Solution

Use a boost regulator module (Figure 8-10). The module shown is from SparkFun, but similar designs are available on eBay more cheaply.

As always with such low-cost purchases from eBay, you should test the module thoroughly before using it. They do not always work exactly as advertised, and quality can be quite variable.

The advantage of this kind of module is that it acts as both a voltage regulator to supply 5V to the Pi and also has a USB socket of its own and a charging circuit. So, if you plug the Pi's normal power adapter into the socket on the charger, the Pi will be powered and the battery charged, allowing you to unplug the USB power and use the Pi on the battery for as long as it has enough power.

With a 1300 mAh LiPo battery, you could expect the Pi to be powered for about two hours.

Figure 8-10. Powering a Raspberry Pi with a 3.7V LiPo battery

Discussion

If you plan to handle the charging of the battery elsewhere, then you can just go for a boost converter module, without the charger, at a lower cost.

See Also

You can also find ready-made USB LiPo battery packs, often with a high-capacity LiPo battery, that will power your Pi for hours.

8.16. Getting Started with a PiFace Digital Interface Board

Problem

You want to know how to set up and use a PiFace interface board.

Solution

The PiFace Digital ("Modules" on page 381) is an expansion board that fits neatly on top of the Raspberry Pi and provides a useful set of inputs and outputs (Figure 8-11).

Figure 8-11. PiFace digital interface board

The PiFace has accompanying software that needs to be set up.

The first step is to enable SPI. To do this, you will need to edit the file */etc/modprobe.d/ raspi-blacklist.conf.* Open this up using nano or your favorite editor and comment out the line blacklisting SPI so that the contents of the file appear:

```
# blacklist spi and i2c by default (many users don't need them)

#blacklist spi-bcm2708
#blacklist i2c-bcm2708
```

You may find that the line for i2c is already commented out.

The next step is to download the installer for the PiFace software. This takes the form of an installer script that will fetch what it needs from the Internet and then install everything for you. To download and run the installer, issue the following commands in the Terminal:

```
$ wget http://pi.cs.man.ac.uk/download/install.txt
$ bash install.txt
```

It takes quite a while for the installation to complete, after which you will need to reboot the Raspberry Pi with this command:

```
$ sudo reboot
```

The easiest way to test out the PiFace is to use the emulator that comes with it (Figure 8-12).

Figure 8-12. PiFace digital emulator

This program allows you to control what is happening to the PiFace attached to the Raspberry Pi, as well as monitor button presses.

To run the emulator, use this command:

```
$ ~/piface/scripts/piface-emulator
```

The emulator has a graphical user interface, so you cannot run it from an SSH session (Recipe 2.7). You need to have keyboard, mouse, and monitor attached to the Raspberry Pi, or use VNC (Recipe 2.8).

Once the emulator is running, click the Override Enable button and then click Output Pin 0. You should hear a little click as one of the two relays closes and at the same time the LED labeled LED0 should light. Repeated clicking of the Output Pin 0 button on the emulator should toggle both the relay and the LED.

The other output buttons will have a similar effect. When you use Output Pin 0 or Output Pin 1, you will see two little blue dots on the righthand side of the emulator.

These will change as you toggle the relay and show the currently closed contacts of the relay.

Similarly, on the emulator, if you select the checkbox labeled "Keep inputs updated" and then press one of the push switches on the left of the actual PiFace board, you will see a little blue dot light up next to the screw terminal for that input.

The emulator is actually a great way of working out where inputs and outputs are located on the board, as well as testing out external electronics attached to the board before you start writing programs to control it.

Discussion

The emulator is a very useful tool for programming the PiFace, but sooner or later, you are going to want to write some actual code.

You can program the PiFace using Python, Scratch, or C. We will use Python. You can start simply by opening the Python console (Recipe 5.3) and entering the following commands:

```
>>> import piface.pfio as pfio
>>> pfio.init()
>>> pfio.digital_write(0, 1)
>>> pfio.digital_write(0, 0)
```

This should allow you to turn LED1 (and its relay) on and off.

The PiFace is equipped with the following:

- Eight digital outputs (open collector)
- Eight digital inputs (3.3V)
- Eight LEDs connected to the outputs
- Four switches connected to inputs 0 to 3
- Two relays connected to outputs 0 and 1

Since the PiFace only has digital inputs and outputs, you only really need two functions:

- The function `digital_write`, which takes two parameters, the first being the output (0 to 7) to write and the second being 1 or 0 for high/low.
- To detect whether an input is high or low, use `digital_read`, which takes the single parameter of the input to check and returns true or false (1 or 0).

To detect a press of switch 1 on the lefthand side of the board, try the following code. You can type it directly into the Python console or use IDLE (Recipe 5.2):

```
>>> import piface.pfio as pfio
>>> import time
```

```
>>> pfio.init()
>>> while True:
...     print(pfio.digital_read(0))
...     time.sleep(1)
...
0
0
1
1
1
1
0
```

When the button is pressed, 1 will be printed out; otherwise, 0 will be printed.

See Also

To use a relay with a Raspberry Pi without an interface board, see Recipe 9.5.

The RaspiRobot board (Recipe 8.18) is another type of interface board for the Raspberry Pi, intended more for controlling robots (Recipe 10.8).

8.17. Getting Started with a Gertboard

Problem

You want to know how to get started with a Gertboard interface board.

Solution

Figure 8-13 shows a Gertboard attached to a Raspberry Pi.

Despite its rather daunting appearance, the Gertboard is pretty easy to get started with. The first step is to install the software examples that accompany the board. The example programs are available for both the RPi.GPIO and Wiring Pi Python libraries. Since we are more familiar with the RPi.GPIO library, we will use those versions. You should install RPi.GPIO if you have not already done so (Recipe 8.3).

To install the Gertboard examples, run the following commands from a Terminal:

```
$ wget http://raspi.tv/download/GB_Python.zip
$ unzip GB_Python.zip
$ cd GB_Python
```

The Gertboard is divided into a number of areas that are disconnected until you join them with the supplied jumper wires. The next example reads the state of the three push buttons built into the Gertboard and writes out a new line each time one of the buttons is pressed or released. As you will see shortly, when you first run the program, it will

Figure 8-13. The Gertboard expansion board

tell you which pins you need to connect together to configure the Gertboard for the example program.

For now, just make the connections according to the instructions; you'll see more about what's going on in the next section:

```
$ sudo python buttons-rg.py
These are the connections for the Gertboard buttons test:
GP25 in J2 --- B1 in J3
GP24 in J2 --- B2 in J3
GP23 in J2 --- B3 in J3
Optionally, if you want the LEDs to reflect button state do the following:
jumper on U3-out-B1
jumper on U3-out-B2
jumper on U3-out-B3
When ready hit enter.

111
110
```

Press the buttons, and you will see how the output changes. Press Ctrl-C when you've had enough experimentation.

Discussion

In this example, you connect GPIO pins 25, 24, and 23 to an area of the board that has a buffer chip with the switches connected to three of its inputs. Its outputs are connected to the GPIO pins, which are configured as inputs. This arrangement of using a buffer protects the inputs of the Raspberry Pi.

The Gertboard has a rich list of features:

- 12 buffered I/O lines
- 3 pushbuttons
- 12 LEDs
- 6 open collector drivers (50V, 0.5A)
- 18V, 2A motor controller
- 28-pin dual in line ATmega microcontroller
- 2-channel 8, 10, or 12 bit digital-to-analog converter
- 2-channel 10 bit analog-to-digital converter

In addition, the ATmega328 has six analog inputs and a whole load of IO lines of its own. The ATmeg328 used is actually the same chip at the heart of the Arduino Uno microcontroller (see Chapter 14). With some effort, you can set it up to be programmed using the Arduino IDE running on the Raspberry Pi to which it is attached (see Recipe 14.15).

See Also

Check out the Gertboard Manual (*http://bit.ly/17FGFLm*).

For some alternative and simpler interface boards, see PiFace (Recipe 8.16) and the RaspiRobot board (Recipe 8.18).

A lot of people prefer to use an Arduino microcontroller board to handle the electronics interfacing, and link it to the Raspberry Pi to do the things that the Raspberry Pi is good at, such as having a large screen or interfacing to networks. This is the subject of Chapter 14.

8.18. Getting Started with a RaspiRobot Board

Problem

You want to know how to use a RaspiRobot board.

Solution

Figure 8-14 shows the RaspiRobot board (version 1). The board has a dual-motor controller that can be used for two DC motors or a single stepper motor. It is also capable of supplying 5V power to the Raspberry Pi using its built-in voltage regulator. The board also has two switch inputs and two low-power outputs, and provides easy connections to the I2C and serial interfaces of the Raspberrry Pi.

Figure 8-14. The RaspiRobot board

After assembling the RaspiRobot board using the instructions supplied, you will need to install the libraries it requires by entering the following commands into a Terminal session on your Raspberry Pi:

```
$ sudo apt-get install python-rpi.gpio
$ sudo apt-get install python-serial
```

To install the accompanying library, issue the following commands in a Terminal window:

```
$ wget https://github.com/simonmonk/raspirobotboard/archive/master.zip
$ unzip master.zip
$ cd raspirobotboard-master
$ sudo python setup.py install
```

Discussion

Fit your RaspiRobot board onto your Raspberry Pi and then power up the Raspberry Pi. You can try out a few commands with the RaspiRobot board using just the Python

console without attaching external power or motors to the RaspiRobot board. Since the RaspiRobot board library uses the GPIO port, you will need to start Python as a super-user using this command:

```
$ sudo python
```

When you first power up, you should find that both LEDs on the RaspiRobot board will be lit. Now enter the following commands, and both LEDs should turn off as the library is initialized:

```
>>> from raspirobotboard import *
>>> rr = RaspiRobot()
```

Try turning one of the LEDs on and off using these commands:

```
>>> rr.set_led1(1)
>>> rr.set_led1(0)
```

The RaspiRobot board has two pairs of pins, each designed to be connected to a switch. These are labeled SW1 and SW2 on the board. You can test whether SW1 is closed using the command:

```
>>> print(rr.sw1_closed())
False
```

Shorting the two pins of SW1 together with a screwdriver and issuing the command again will return True.

Other commands are available to set the two open collector outputs (set_oc1 and set_oc2) as well as motor control commands (forward, reverse, left, right, and stop). See *http://www.raspirobot.com* for the full command reference.

Of course, you can use the RaspiRobot Board interface with just the RPi.GPIO library. The GPIO pins used by the RaspiRobot Board are listed in Table 8-1.

Table 8-1. RaspiRobot board GPIO pin usage

Motor 1A	17
Motor 1B	4
Motor 2A	10
Motor 2B	25
SW1	11
SW2	9
OC1	22
OC2	27

Figure 8-15. The Humble Pi

See Also

You can find out more about the RaspiRobot board and other projects that use it at the RaspiRobot website (*http://www.raspirobot.com*).

For a recipe that uses this board to make a roving robot, see Recipe 10.8.

To use the RaspiRobot board to control a bipolar stepper motor, see Recipe 10.7.

8.19. Using a Humble Pi Prototyping Board

Problem

You want to know how to use a Humble Pi prototyping board.

Solution

The Humble Pi (Figure 8-15) is a prototyping board rather than an interface board like the PiFace (Recipe 8.16) or RaspiRobot board (Recipe 10.8). In other words, it does not include any electronics—it is designed for you to solder your own components to a prototyping area.

A neat feature of the Humble Pi is that it has a special area of the board where you can attach a DC barrel socket, voltage regulator, and associated capacitors. Ciseco (*http://*

www.ciseco.co.uk), the maker of the board, also sells a suitable regulator and capacitor kit.

Discussion

The board design has a grid of holes at a pitch of the standard 0.1 inch used by most through-hole components, including DIL ICs. You solder components to the board by pushing the leads through the holes at the top and soldering the connections underneath.

If you have not soldered before, check out a few YouTube videos to get the basic idea.

Holes contained within a white rectangle are connected together by tracks on the PCB, and the board is organized with central power lines, one connected to GND and one that can be bridged to 5V or 3.3V on the Raspberry Pi or to the output of the optional voltage regulator.

On either side of these central power lines are rows comprising groups of three connections. This gives plenty of room for DIL ICs, either accross the central power lines or on either side.

Having soldered the components into place, you will need extra wires to link everything. These can be run on the bottom of the board or the top, or both if the design is complicated.

In any case, it is a good idea to plan out the layout before you start soldering.

See Also

There is more information about this product at the Ciseco website (*http://shop.cise co.co.uk/*).

The Pi Plate from Adafruit is a similar design to the Humble Pi (Recipe 8.20).

8.20. Using a Pi Plate Prototyping Board

Problem

You want to know how to use a Pi Plate prototyping board.

Solution

The Pi Plate (Figure 8-16) is a prototyping board rather than an interface board like the PiFace (Recipe 8.16) or RaspiRobot board (Recipe 10.8). In other words, it does not include any electronics—it is designed for you to solder your own components to a prototyping area.

Figure 8-16. The Pi Plate

Unlike the Humble Pi (Recipe 8.19), the Pi Plate does not have cutouts to work around the taller sockets on the Raspberry Pi; it is rectangular and exactly the same size as the Pi itself. It achieves this by using an extra tall header socket that lifts the whole board high enough to clear all the components on the Raspberry Pi.

The board also has an area where a 16-pin, surface-mount chip can be used and a row of four holes, spaced more widely, where a screw terminal can be soldered in place.

The board has screw terminals on two sides that are connected to all the GPIO pins. You can ignore the prototyping area entirely and just use the screw terminals to attach things directly to the GPIO pins without soldering.

Discussion

The board design has a grid of holes at a pitch of the standard 0.1 inch used by most through-hole components, including DIL ICs. You solder components to the board by pushing the leads through the holes at the top and soldering the connections underneath.

The tracks that connect the holes are clearly visible on the top of the board, and the board is divided into several zones. There is an area with central power busses intended for DIL ICs as well as a general prototyping area and areas for a surface-mount chip and extra screw terminals.

Having soldered the components into place, you will need extra wires to link everything up. These can be run on the bottom of the board or the top, or both if the design is complicated.

It's a good idea to plan out the layout before you start soldering.

In the following instructions, you will build an RGB LED onto a Pi Plate using the layout shown in Figure 8-17. This recipe is a version of Recipe 9.9, which is the same except that the design is built onto a solderless breadboard.

The first step is to solder the resistors into place. Bend the leads and push them through the appropriate holes in the board. Then, flip the board over and touch the soldering iron to where the leads emerge from the hole for a second or so before applying the solder, which should flow around the lead (Figure 8-18).

When you have soldered both ends, snip off the excess lead and repeat for the other two resistors (Figure 8-19).

Now solder the LED into place, taking care to get it the right way around. The longest lead is the common cathode and should be the only LED lead attached to the strip of holes that is not connected to any resistor. Very occasionally, you will find LEDs where the longer lead is not the positive lead. This is often the case with infrared LEDs. So, if you aren't sure, check the datasheet for the LED or the supplier's information page.

You will also need to solder a short length of wire from that row to the GND connection on the Humble Pi (Figure 8-20).

When the board is complete, it should look like Figure 8-21.

You can try out the LED using the Python program from Recipe 9.9.

See Also

There is more information about this product at the Adafruit website (*http://www.adafruit.com/products/801*).

The Pi Plate from Adafruit is a similar design to the Humble Pi (Recipe 8.20). The example in that recipe will give you an idea of how to solder components onto the Humble Pi.

8.21. Using a Paddle Terminal Breakout Board

Problem

You want to connect to a GPIO breakout board with paddle terminals.

Figure 8-17. Board layout for an RGB LED on a Pi Plate

Figure 8-18. Soldering a resistor to the Pi Plate

Figure 8-19. Resistors soldered to the Pi Plate

Figure 8-20. Soldering the link wire to the Pi Plate

Figure 8-21. The finished RGB LED on a Pi Plate

Solution

You can use the paddle terminal breakout board shown in Figure 8-22 to quickly connect bare wires or component leads to the Raspberry Pi without having to solder. You simply push the spring-loaded slide down, put a wire or component lead into the hole, and release the slide.

Figure 8-22. A paddle terminal breakout board

Discussion

One of these devices is occasionally useful for a quick wireup.

See Also

An alternative is to use the screw terminals provided in a Pi Plate (Recipe 8.20).

Controlling Hardware

9.0. Introduction

In this chapter, you come to grips with the control of electronics through the Raspberry Pi's GPIO connector.

Most of the recipes require the use of solderless breadboard and male-to-female and male-to-male jumper wires (see Recipe 8.10).

9.1. Connecting an LED

 Be sure to check out the accompanying video for this recipe at *http://razzpisampler.oreilly.com*.

Problem

You want to know how to connect an LED to the Raspberry Pi.

Solution

Connect an LED (see "Opto-Electronics" on page 381) to one of the GPIO pins using a 470Ω or 1kΩ series resistor (see "Resistors and Capacitors" on page 380) to limit the current. To make this recipe, you will need:

- Breadboard and jumper wires (see "Prototyping Equipment" on page 380)
- 1kΩ resistor (see "Resistors and Capacitors" on page 380)
- LED (see "Opto-Electronics" on page 381)

Figure 9-1 shows how you can wire this using solderless breadboard and male-to-female jumper leads.

Figure 9-1. Connecting an LED to a Raspberry Pi

Having connected the LED, we need to be able to turn it on and off using commands from Python. To do this, follow Recipe 8.3 to install the RPi.GPIO Python library.

Start a Python console (Recipe 5.3) from the Terminal with superuser access and enter these commands:

```
$ sudo python
>>> import RPi.GPIO as GPIO
>>> GPIO.setmode(GPIO.BCM)
>>> GPIO.setup(18, GPIO.OUT)
>>> GPIO.output(18, True)
>>> GPIO.output(18, False)
```

This will turn your LED on and off.

Discussion

LEDs are a very useful, cheap, and efficient way of producing light, but you do have to be careful how you use them. If they are connected directly to a voltage source (such as a GPIO output) that is greater than about 1.7 volts, they will draw a very large current. This can often be enough to either destroy the LED or whatever is providing the current —which is not good if your Raspberry Pi is providing the current.

You should always use a series resistor with an LED because the series resistor is placed *between* the LED and the voltage source, which limits the amount of current flowing through the LED to a level that is safe for both the LED and the GPIO pin driving it.

Raspberry Pi GPIO pins can only provide about 3 mA of current. LEDs will generally illuminate with any current greater than 1 mA, but will be brighter with more current. Use Table 9-1 as a guide to selecting a series resistor based on the type of LED; the table also indicates the approximate current that will be drawn from the GPIO pin.

Table 9-1. Selecting series resistors for LEDs and a 3.3V GPIO pin

LED type	Resistor	Current (mA)
Red	470Ω	3.5
Red	1kΩ	1.5
Orange, yellow, green	470Ω	2
Orange, yellow, green	1kΩ	1
Blue, white	100Ω	3
Blue, white	270Ω	1

As you can see, in all cases, it is safe to use a 470Ω resistor. If you are using a blue or white LED, you can reduce the value of the series resistor considerably. If you want to play it safe, use 1kΩ.

If you wanted to extend the experiments that you made in the Python console into a program that makes the LED blink on and off repeatedly, you could paste the following code into the IDLE (Recipe 5.2) or nano (Recipe 3.6) editors. Save the file as *led_blink.py*. You can also download the program from the *Raspberry Pi Cookbook* website (*http://www.raspberrypicookbook.com*). Follow the link to this book and then the Downloads section.

```
import RPi.GPIO as GPIO
import time

GPIO.setmode(GPIO.BCM)
GPIO.setup(18, GPIO.OUT)

while (True):
    GPIO.output(18, True)
    time.sleep(0.5)
        GPIO.output(18, False)
        time.sleep(0.5)
```

Remember that to run the program, you must have superuser privileges for the RPi.GPIO library, so you need to use this command:

```
$ sudo python led_blink.py
```

See Also

Check out this handy series resistor calculator (*http://led.linear1.org/1led.wiz*).

For more information on using breadboard and jumper wires with the Raspberry Pi, see Recipe 8.10.

9.2. Controlling the Brightness of an LED

Problem

You want to vary the brightness of an LED from a Python program.

Solution

The RPi.GPIO library has a pulse-width modulation (PWM) feature that allows you to control the power to an LED and its brightness.

To try it out, connect an LED as described in Recipe 9.1 and run this test program:

```
import RPi.GPIO as GPIO

led_pin = 18
GPIO.setmode(GPIO.BCM)
GPIO.setup(led_pin, GPIO.OUT)

pwm_led = GPIO.PWM(led_pin, 500)
pwm_led.start(100)

while True:
        duty_s = raw_input("Enter Brightness (0 to 100):")
        duty = int(duty_s)
        pwm_led.ChangeDutyCycle(duty)
```

If you are using Python 3 rather than Python 2, change the command `raw_input` to just `input`.

Run the Python program, and you will be able to change the brightness by entering a number between 0 and 100:

```
pi@raspberrypi ~ $ sudo python led_brightness.py
Enter Brightness (0 to 100):0
Enter Brightness (0 to 100):20
Enter Brightness (0 to 100):10
Enter Brightness (0 to 100):5
Enter Brightness (0 to 100):1
Enter Brightness (0 to 100):90
```

You can exit the program by pressing Ctrl-C.

Discussion

PWM is a clever technique where you vary the length of pulses while keeping the overall number of pulses per second (the frequency in Hz) constant. Figure 9-2 illustrates the basic principle of PWM.

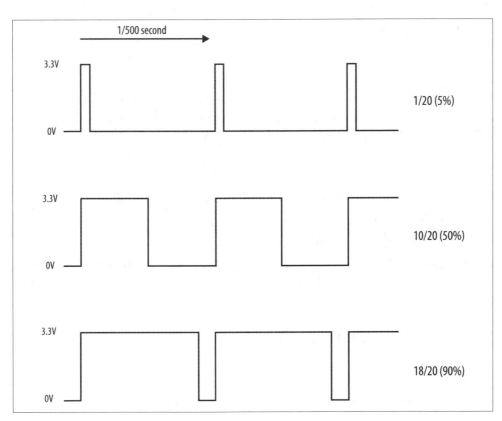

Figure 9-2. Pulse-width modulation

At high frequencies, the measured PWM frequency varies somewhat from the frequency supplied as an argument. This may be something that changes in later versions of the PWM feature of RPi.GPIO.

You can change the PWM frequency by modifying this line:

```
pwm_led = GPIO.PWM(led_pin, 500)
```

The value is in Hz, so in this case, the frequency is set to 500 Hz.

Table 9-2 compares frequencies specified in the second parameter to GPIO.PWM to the actual frequency on the pin measured with an oscilloscope.

Table 9-2. Requested frequency against actual frequency

Requested frequency	Measured frequency
50 Hz	50 Hz
100 Hz	98.7 Hz
200 Hz	195 Hz
500 Hz	470 Hz
1 kHz	890 Hz
10 kHz	4.4 kHz

I also found that as the frequency increased, its stability decreased. This means that this PWM feature is no good for audio, but plenty fast enough for controlling the brightness of LEDs or the speed of motors.

See Also

For more information on PWM, see Wikipedia (*http://bit.ly/1iobPt8*).

Recipe 9.9 uses PWM to change the color of an RGB LED, and Recipe 10.3 uses PWM to control the speed of a DC motor.

For more information on using breadboard and jumper wires with the Raspberry Pi, see Recipe 8.10. You can also control the brightness of the LED with a slider control—see Recipe 9.8.

9.3. Make a Buzzing Sound

Problem

You want to make a buzzing sound with the Raspberry Pi.

Solution

Use a piezo-electric buzzer connected to a GPIO pin.

Most small piezo buzzers work just fine using the arrangement shown in Figure 9-3. The one I used is an Adafruit supplied component (see "Miscellaneous" on page 382). You can connect the buzzer pins directly to the Raspberry Pi using female-to-female headers (see "Prototyping Equipment" on page 380).

These buzzers use very little current. However, if you have a large buzzer or just want to play it safe, then put a 470Ω resistor between the GPIO pin and the buzzer lead.

Figure 9-3. Connecting a piezo buzzer to a Raspberry Pi

Paste the following code into the IDLE (Recipe 5.2) or nano (Recipe 3.6) editors. Save the file as *buzzer.py*. You can also download the program from the Downloads section of the *Raspberry Pi Cookbook* website (*http://www.raspberrypicookbook.com*).

```python
import RPi.GPIO as GPIO
import time

buzzer_pin = 18
GPIO.setmode(GPIO.BCM)
GPIO.setup(buzzer_pin, GPIO.OUT)

def buzz(pitch, duration):
    period = 1.0 / pitch
    delay = period / 2
    cycles = int(duration * pitch)
    for i in range(cycles):
        GPIO.output(buzzer_pin, True)
        time.sleep(delay)
        GPIO.output(buzzer_pin, False)
        time.sleep(delay)

while True:
    pitch_s = raw_input("Enter Pitch (200 to 2000): ")
    pitch = float(pitch_s)
    duration_s = raw_input("Enter Duration (seconds): ")
```

```
        duration = float(duration_s)
        buzz(pitch, duration)
```

When you run the program, it will first prompt you for the pitch in Hz and then the duration of the buzz in seconds:

```
$ sudo python buzzer.py
Enter Pitch (2000 to 10000): 2000
Enter Duration (seconds): 20
```

Discussion

Piezo buzzers don't have a wide range of frequencies, nor is the sound quality remotely good. However, you can vary the pitch a little. The frequency generated by the code is very approximate.

The program works by simply toggling the GPIO pin 18 on and off with a short delay in between. The delay is calculated from the pitch. The higher the pitch (frequency), the shorter the delay needs to be.

See Also

You can find the datasheet for the piezo buzzer here: *http://bit.ly/Iwkv2R*.

9.4. Switching a High-Power DC Device Using a Transistor

Problem

You want to control the power to a high-power, low-voltage DC device such as a 12V LED module.

Solution

These high-power LEDs use far too much current to light directly from a GPIO pin. They also require 12V rather than the 3.3V. To control such a high-power load, you need to use a transistor.

In this case, you will use a high-power type of transistor called a MOSFET (metal oxide semiconductor field effect transistor), which costs less than a dollar but can handle loads up to 30 amps, many times more than is required for the high-power LEDs. The MOS-FET used is a FQP30N06 (see "Transistors and Diodes" on page 381).

Figure 9-4 shows how you can connect a MOSFET on a breadboard. Make sure that you correctly identify the positive and negative supply leads for the LED module.

Figure 9-4. Controlling large currents with a MOSFET

To make this recipe, you will need:

- Breadboard and jumper wires (see "Prototyping Equipment" on page 380)
- 1kΩ resistor (see "Resistors and Capacitors" on page 380)
- FQP30N06 N-Channel MOSFET (see "Transistors and Diodes" on page 381)
- 12V power adapter
- 12V DC LED module

The Python code to turn the LED panel on and off is exactly the same as if we were controlling a single low-power LED without the MOSFET (see Recipe 9.2).

You can also use PWM with the MOSFET to control the brightness of the LED module (see Recipe 9.2).

Discussion

Whenever you need to power anything significant using the GPIO connector, use batteries or an external power adapter. The GPIO connector can only supply relatively low currents (Recipe 8.2). In this case, you'll use a 12V DC power adapter to provide the power to the LED panel. Pick a power adapter that has sufficient power handling. So, if the LED module is 5W, then you need at least a 12V 5W power supply (6W would be better). If the power supply specifies a maximum current rather than power, then you

can calculate its power by multiplying the voltage by the maximum current. So, a 500 mA 12V power supply can provide 6 W of power.

The resistor is necessary to ensure that the peak currents that occur as the MOSFET switches from off to on and vice versa do not overload the GPIO pin. The MOSFET switches the negative side of the LED panel, so the positive supply is connected directly to the positive side of the LED panel, and the negative side of the LED panel is connected to the *drain* of the MOSFET. The *source* connection of the MOSFET is connected to GND, and the MOSFET's *gate* pin controls the flow of current from the drain to the source. If gate voltage is above about 2V, the MOSFET will turn on and current flows through both it and the LED module.

This circuit is suitable for controlling the power to other low-voltage DC devices. The only real exceptions are motors and relays, which require some extra treatment (see Recipe 10.3).

See Also

Check out the datasheet for the MOSFET (*http://bit.ly/18J3bxT*).

If you would like to create a graphical user interface with which to control your LED module, see Recipe 9.7 for simple on/off control and Recipe 9.8 for variable control of the brightness with a slider.

9.5. Switching a High-Power Device Using a Relay

Problem

You want to turn devices on and off that may not be suitable for switching with a MOSFET.

Solution

Use a relay and small transistor.

Figure 9-5 shows how you can connect up a transistor and relay on breadboard. Make sure that both the transistor and diode are placed the right way around. The diode has a stripe at one end, and the transistor used here has one side flat and one curved.

To make this recipe, you will need:

- Breadboard and jumper wires (see "Prototyping Equipment" on page 380)
- 1kΩ resistor (see "Resistors and Capacitors" on page 380)
- Transistor 2N3904 (see "Transistors and Diodes" on page 381)
- 1N4001 diode (see "Transistors and Diodes" on page 381)

Figure 9-5. Using a relay with a Raspberry Pi

- 5V relay (see "Miscellaneous" on page 382)
- Multimeter

You can use the same LED blink program that you used in Recipe 9.1. If all is well, you'll hear a click from the relay each time the contacts are closed. However, relays are slow mechanical devices, so don't try to use them with PWM. It may damage the relay.

Discussion

Relays have been around since the early days of electronics and have the great advantage of being easy to use, plus they'll work in any situation where a switch would normally work—for example, when you're switching AC (alternating current) or in situations where the exact wiring of the device being switched is unknown.

If the relay contacts are asked to exceed their specifications, then the relay's life will be shortened. There will be arcing, and the contacts may eventually fuse together. There is

Figure 9-6. The workings of a relay

also the possibility of the relay becoming dangerously hot. When in doubt, overspecify the relay contacts.

Figure 9-6 shows how a relay is constructed.

A relay is essentially a switch whose contacts are closed when an electromagnet pulls them together. Since the electromagnet and switch are not connected electrically in any way, this protects the circuit driving the relay coil from any high voltages on the switch side.

The downside of relays is that they are slow to operate and will eventually wear out after many hundreds of thousands of operations. This means they are only suitable for slow on/off control, not for fast switching like PWM.

The coil of a relay requires about 50 mA to close the connections. Because a Raspberry Pi GPIO pin is only capable of supplying about 3 mA, you need to use a small transistor as a switch. You don't need to use a high-power MOSFET like you did in Recipe 9.4, but instead you can just use a small transistor. This has three connections. The base (middle lead) is connected to the GPIO pin via a 1kΩ resistor to limit the current. The *emitter* is connected to GND, and the *collector* to one side of the relay. The other side of the relay is connected to 5V on the GPIO connector. The diode is used to suppress any high-voltage pulses that occur when the transistor rapidly switches the power to the relay's coil.

 Although relays can be used to switch 110V AC, this voltage is very dangerous and should not be used on a breadboard. If you want to switch high voltages, use Recipe 9.6 instead.

See Also

For switching DC using a power MOSFET, see Recipe 9.2.

Another way of controlling a relay is to use a PiFace interface board (Recipe 8.16).

9.6. Controlling High-Voltage AC Devices

Problem

You want to switch 110 or 240V AC using a Raspberry Pi.

Solution

Use a PowerSwitch Tail II (see "Modules" on page 381). This handy device makes it really easy to switch AC equipment on and off from a Raspberry Pi. It has an AC socket on one end and a plug on the other, like an extension cable; the only difference is that the control box in the middle of the lead has three screw terminals. By attaching terminal 2 to GND and terminal 1 to a GPIO pin, the device acts like a switch to turn the appliance on and off.

You can use the same Python code as you did in Recipe 9.1 to use the PowerSwitch Tail, as shown in Figure 9-7.

Discussion

The PowerSwitch Tail uses a relay, but to switch the relay, it uses a component called an *opto-isolator* that has an LED shining onto a photo-TRIAC (a high-voltage, light-sensitive switch); when the LED is illuminated, the photo-TRIAC conducts, supplying current to the relay coil.

The LED inside the opto-isolator has its current limited by a resistor, so that only 3 mA flows through it when you supply it with 3.3V from a GPIO pin.

See Also

For switching DC using a power MOSFET, see Recipe 9.7, and for switching using a relay on breadboard, see Recipe 9.5.

A 240V version of the PowerSwitch Tail is available as a kit (*http://bit.ly/1byrZtl*).

9.7. Making a User Interface to Turn Things On and Off

Problem

You want to make an application to run on the Raspberry Pi that has a button for turning things on and off.

Figure 9-7. Using a PowerSwitch Tail with Raspberry Pi

Solution

Using the Tkinter user interface framework, write a Python program that uses a check-button to turn the GPIO pin on and off (Figure 9-8).

Figure 9-8. A user interface for turning things on and off

You'll need to connect an LED or some other kind of output device to GPIO pin 18. Using an LED (Recipe 9.1) is the easiest option to start with.

Open an editor (nano or IDLE) and paste in the following code. As with all the program examples in this book, you can also download the program from the Code section of the *http://www.raspberrypicookbook.com*, where it is called *gui_switch.py*.

```
from Tkinter import *
import RPi.GPIO as GPIO
import time

GPIO.setmode(GPIO.BCM)
GPIO.setup(18, GPIO.OUT)

class App:

    def __init__(self, master):
        frame = Frame(master)
        frame.pack()
        self.check_var = BooleanVar()
        check = Checkbutton(frame, text='Pin 18',
                command=self.update,
                variable=self.check_var, onvalue=True, offvalue=False)
        check.grid(row=1)

    def update(self):
        GPIO.output(18, self.check_var.get())

root = Tk()
root.wm_title('On / Off Switch')
app = App(root)
root.geometry("200x50+0+0")
root.mainloop()
```

Note that you will need to run it with sudo because the RPi.GPIO requires you to have superuser privileges to access the GPIO hardware:

```
$ sudo python gui_switch.py
```

 In Python 3, the Tkinter library has been renamed *tkinter* with a low-ercase *t*.

Discussion

The example program defines a class called App that contains most of the application code. Its initializer function creates a member variable called check_var that contains an instance of BooleanVar that is then supplied as the variable option to the check-button. This ensures that every time the checkbutton is clicked, the value in this variable will be changed. The command option runs the update command every time such a change occurs.

The update function simply writes the value in check_var to the GPIO output.

See Also

You can use this program to control an LED (Recipe 9.1), a high-power DC device (Recipe 9.4), a relay (Recipe 9.5), or high-voltage AC device (Recipe 9.6).

9.8. Making a User Interface to Control PWM Power for LEDs and Motors

Problem

You want to make an application to run on the Raspberry Pi that has a slider to control power to a device using PWM.

Solution

Using the Tkinter user interface framework, write a Python program that uses a slider to change the PWM duty cycle between 0 and 100 percent (Figure 9-9).

Figure 9-9. User interface for controlling PWM power

You will need to connect an LED or some other kind of output device to GPIO pin 18 that is capable of responding to a PWM signal. Using an LED (Recipe 9.1) is the easiest option to start with.

Open an editor (nano or IDLE) and paste in the following code. As with all the program examples in this book, you can also download the program from the Code section of *http://www.raspberrypicookbook.com*, where it is called *gui_slider.py*.

```
from Tkinter import *
import RPi.GPIO as GPIO
import time

GPIO.setmode(GPIO.BCM)
GPIO.setup(18, GPIO.OUT)
pwm = GPIO.PWM(18, 500)
pwm.start(100)

class App:
```

```
    def __init__(self, master):
        frame = Frame(master)
        frame.pack()
        scale = Scale(frame, from_=0, to=100,
                orient=HORIZONTAL, command=self.update)
        scale.grid(row=0)

    def update(self, duty):
        pwm.ChangeDutyCycle(float(duty))

root = Tk()
root.wm_title('PWM Power Control')
app = App(root)
root.geometry("200x50+0+0")
root.mainloop()
```

Note that you will need to run it with sudo because the RPi.GPIO requires you to have superuser privileges to access the GPIO hardware.

```
$ sudo python gui_slider.py
```

 In Python 3, the Tkinter library has been renamed *tkinter*, with a lowercase *t*.

Discussion

The example program defines a class called App that contains most of the application code. The command option runs the update command every time the value of the slider is changed. This updates the duty cycle of the output pin.

See Also

You can use this program to control an LED (Recipe 9.1), a DC motor (Recipe 10.3), or a high-power DC device (Recipe 9.4).

9.9. Changing the Color of an RGB LED

Problem

You want to control the color of an RGB LED.

Solution

Use PWM to control the power to each of the red, green, and blue channels of an RGB LED.

To make this recipe, you will need:

- Breadboard and jumper wires (see "Prototyping Equipment" on page 380)
- Three 1kΩ resistors (see "Resistors and Capacitors" on page 380)
- RGB common cathode LED ("Opto-Electronics" on page 381)
- A Pi Plate (see Recipe 8.20) or Humble Pi (Recipe 8.19) to make a more permanent project (Optional)

Figure 9-10 shows how you can connect your RBG LED on a breadboard. Make sure that the LED is the correct way around; the longest lead should be the second lead from the top of the breadboard. This connection is called the *common cathode*, as the negative connections (cathodes) of the red, green, and blue LEDs within the LED case have all their negative sides connected together to reduce the number of pins needed in the package.

Figure 9-10. Using a RGB LED with a Raspberry Pi

If you want to make a more permanent design, you can solder this up on a Pi Plate or Humble Pi. See Recipe 8.19 to learn how.

The upcoming program has three sliders to control the red, green, and blue channels of the LED (Figure 9-11).

Figure 9-11. Using a user interface to control an RGB LED

Open an editor (nano or IDLE) and paste in the following code. As with all the program examples in this book, you can also download the program from the Code section of *http://www.raspberrypicookbook.com*, where it is called *gui_sliderRGB.py*.

```
from Tkinter import *
import RPi.GPIO as GPIO
import time

GPIO.setmode(GPIO.BCM)
GPIO.setup(18, GPIO.OUT)
GPIO.setup(23, GPIO.OUT)
GPIO.setup(24, GPIO.OUT)

pwmRed = GPIO.PWM(18, 500)
pwmRed.start(100)

pwmGreen = GPIO.PWM(23, 500)
pwmGreen.start(100)

pwmBlue = GPIO.PWM(24, 500)
pwmBlue.start(100)

class App:

    def __init__(self, master):
        frame = Frame(master)
        frame.pack()
```

```
Label(frame, text='Red').grid(row=0, column=0)
Label(frame, text='Green').grid(row=1, column=0)
Label(frame, text='Blue').grid(row=2, column=0)
scaleRed = Scale(frame, from_=0, to=100,
        orient=HORIZONTAL, command=self.updateRed)
scaleRed.grid(row=0, column=1)
scaleGreen = Scale(frame, from_=0, to=100,
        orient=HORIZONTAL, command=self.updateGreen)
scaleGreen.grid(row=1, column=1)
scaleBlue = Scale(frame, from_=0, to=100,
        orient=HORIZONTAL, command=self.updateBlue)
scaleBlue.grid(row=2, column=1)

    def updateRed(self, duty):
        pwmRed.ChangeDutyCycle(float(duty))

    def updateGreen(self, duty):
        pwmGreen.ChangeDutyCycle(float(duty))

    def updateBlue(self, duty):
        pwmBlue.ChangeDutyCycle(float(duty))

root = Tk()
root.wm_title('RGB LED Control')
app = App(root)
root.geometry("200x150+0+0")
root.mainloop()
```

Discussion

The code is similar in operation to the control for a single PWM channel, described in Recipe 9.8. However, in this case, you need three PWM channels and three sliders, one for each color.

The type of RGB LED used here is a common cathode. If you have the common anode type, then you can still use it, but connect the common anode to the 3.3V pin on the GPIO connector. You will then find that the slider becomes reversed, so a setting of 100 becomes *off* and 0 becomes full *on*.

When you are selecting an LED for this project, LEDs labeled *diffused* are best because it allows the colors to be mixed better.

See Also

If you just want to control one PWM channel, see Recipe 9.8.

For more information on using a breadboard and jumper wires with the Raspberry Pi, see Recipe 8.10.

9.10. Using Lots of LEDs (Charlieplexing)

 Be sure to check out the accompanying video for this recipe at *http://razzpisampler.oreilly.com*.

Problem

You want to control lots of LEDs using as few GPIO pins as possible.

Solution

The way to do this is to use a technique called *Charlieplexing*. The name comes from the inventor, Charlie Allen of the company Maxim, and the technique takes advantage of the feature of GPIO pins that allows them to be changed from outputs to inputs while a program is running. When a pin is changed to be an input, not enough current will flow through it to light an LED or influence other pins connected to the LED that are set as outputs.

Figure 9-12 shows the arrangement for controlling six LEDs with three pins.

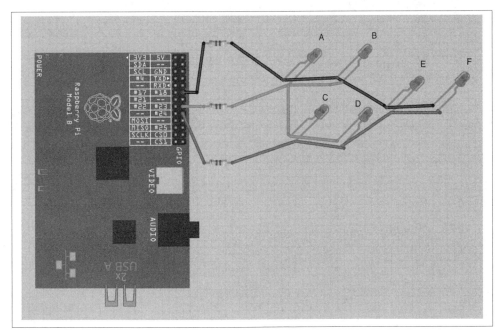

Figure 9-12. Charlieplexing

Figure 9-13 shows the breadboard layout for the LEDs and resistors.

Figure 9-13. Charlieplexing breadboard layout

To make this recipe, you will need:

- Breadboard and jumper wires (see "Prototyping Equipment" on page 380)
- Three 470Ω resistors (see "Resistors and Capacitors" on page 380)
- Six LEDs (see "Opto-Electronics" on page 381)

Open an editor (nano or IDLE) and paste in the following code. As with all the program examples in this book, you can also download the program from the Code section of the *Raspberry Pi Cookbook website* (*http://www.raspberrypicookbook.com*), where it is called *charlieplexing.py*.

This example code prompts you to enter a number between 0 and 5 and then lights one of the six LEDs:

```
import RPi.GPIO as GPIO

pins = [18, 23, 24]

pin_led_states = [
  [1, 0, -1], # A
  [0, 1, -1], # B
  [-1, 1, 0], # C
  [-1, 0, 1], # D
  [1, -1, 0], # E
  [0, -1, 1]  # F
]
```

```
GPIO.setmode(GPIO.BCM)

def set_pin(pin_index, pin_state):
    if pin_state == -1:
        GPIO.setup(pins[pin_index], GPIO.IN)
    else:
        GPIO.setup(pins[pin_index], GPIO.OUT)
        GPIO.output(pins[pin_index], pin_state)

def light_led(led_number):
    for pin_index, pin_state in enumerate(pin_led_states[led_number]):
        set_pin(pin_index, pin_state)

set_pin(0, -1)
set_pin(1, -1)
set_pin(2, -1)

while True:
    x = int(raw_input("Pin (0 to 5):"))
    light_led(x)
```

Discussion

To understand how Charlieplexing works, imagine that you want to light LED A in Figure 9-12. An LED will only light when its positive lead is high, and its negative lead is low. If the voltage is the other way around, it will not light. To light LED A, you need its lead connected to GPIO 18 (via a resistor) to be high and the other lead to LED A, connected to GPIO 23 by a resistor, to be low. However, you must also make sure that GPIO 24 is set to be an input; otherwise, LED C or D will also light depending on whether GPIO 24 is high or low.

The array *pin_led_states* holds the settings for each GPIO for each of the six LEDs. If the value is 0, the pin is low; 1 means high and -1 means set to be an input.

The number of LEDs that can be controlled per GPIO pin is given by the formula:

LEDs = n^2 - n

Using 4 pins, we can have 16, 4, or 12 LEDs, whereas 10 pins would give you a massive 90 LEDs.

In this example, you're lighting only one LED at a time. To light more than one at a time, you need to run a refresh loop that keeps the desired state of the LEDs in an array and refreshes the display, turning on the LEDs that need to be on before moving on to the next. It must do this sufficiently fast so that it appears that more than one of the LEDs is on at the same time.

The more LEDs you use when it comes to making it appear that more than one LED is on at a time, the less time the LED will actually be lit, and the dimmer the LEDs will become.

See Also

For more information about Charlieplexing, see Wikipedia (*http://bit.ly/1aBpKaW*). If you are interested in using only a single LED, see Recipe 9.1.

9.11. Using an Analog Meter as a Display

Problem

You want to connect an analog panel volt meter to a Raspberry Pi.

Solution

Assuming you have a 5V volt meter, you can use a PWM output to drive the meter directly, connecting the negative side of the meter to ground and the positive side to a GPIO pin (Figure 9-14). However, if the meter is the common 5V kind, you'll only be able to display voltages up to 3.3V.

If you want to use almost the full range of a 5V volt meter, you will need a transistor to act as a switch for the PWM signal and a 1kΩ resistor to limit the current to the base of the transistor.

To make this recipe, you will need:

- 5V panel meter ("Miscellaneous" on page 382)
- Breadboard and jumper wires (see "Prototyping Equipment" on page 380)
- Two 1kΩ resistors (see "Resistors and Capacitors" on page 380)
- Transistor 2N3904 (see "Transistors and Diodes" on page 381)

The breadboard layout for this is shown in Figure 9-15.

Discussion

To test the volt meter, use the same program as you did for controlling the brightness of the LED in Recipe 9.2.

You will probably notice that the needle gives a steady reading at either end of the scale, but everywhere else jitters a bit. This is a side effect of the way the PWM signals are generated. For a steadier result, you can use external PWM hardware like the 16-channel module used in Recipe 10.2.

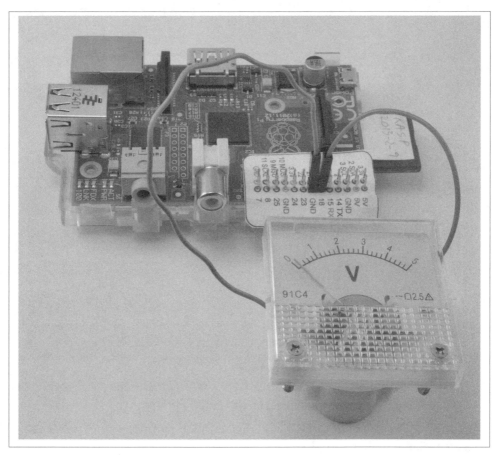

Figure 9-14. Connecting a volt meter directly to a GPIO pin

See Also

For more information about how old-fashioned volt meters work, see Wikipedia (*https://en.wikipedia.org/wiki/Voltmeter*).

For more information on using a breadboard and jumper wires with the Raspberry Pi, see Recipe 8.10.

9.12. Programming with Interrupts

Problem

You want to respond to some event, such as a button push, without having to continually poll the input pin to see if its state has changed.

Figure 9-15. Using a 5V panel meter with 3.3V GPIO

Solution

Use the `add_event_detect` function of the RPi.GPIO library.

The upcoming example shows how you can attach an interrupt service routine to be triggered when a button is pressed.

Wire the switch onto a breadboard, as shown in Figure 9-16.

Open an editor (nano or IDLE) and paste in the following code. As with all the program examples in this book, you can also download the program from the Code section of the *Raspberry Pi Cookbook* website (*http://www.raspberrypicookbook.com*), where it is called *interrupts.py*.

This example code continually updates a count in seconds and displays a message when the button is pressed:

```
import RPi.GPIO as GPIO
import time

GPIO.setmode(GPIO.BCM)

def my_callback(channel):
    print('You pressed the button')
```

Figure 9-16. Connecting a switch to a GPIO input to demonstrate interrupts

```
GPIO.setup(18, GPIO.IN, pull_up_down=GPIO.PUD_UP)
GPIO.add_event_detect(18, GPIO.FALLING, callback=my_callback)

i = 0
while True:
    i = i + 1
    print(i)
    time.sleep(1)
```

Try running the program with superuser privileges—you should see something like this happen when you press the button:

```
$ sudo python interrupts.py
1
2
3
You pressed the button
4
You pressed the button
5
You pressed the button
```

```
You pressed the button
6
```

Discussion

You could detect when a button has been pressed, or a GPIO input has changed, by simply checking repeatedly in a loop; for example:

```
while True:
    if GPIO.input(18) == False:
        # put the code to be actioned here
    time.sleep(0.1)
```

The disadvantage here is that you can't do much else while you are checking for button presses. A second disadvantage is that if the button press is very quick, it could come and go before you can register it with the GPIO.input. This approach is called *polling*.

Interrupts work differently. They allow you to associate a function with one of the pins so that when the voltage at the input changes either from low to high or vice versa, you can trigger the function to be run.

You can see how this works in the preceding example program. First, define a function called my_callback that takes a single argument. This argument specifies the input that triggered the interrupt, allowing you to use the same handler function for a number of interrupts.

```
def my_callback(channel):
    print('You pressed the button')
```

In this case, the callback function just displays a message.

The line of code that does the actual linking is:

```
GPIO.add_event_detect(18, GPIO.FALLING, callback=my_callback)
```

The first parameter specifies the pin (18). The second can be GPIO.FALLING or GPIO.RISING. If this is set to FALLING, the function will only be called if the GPIO pin goes from high to low. This is the case in this example, as the switch pulls the input low against the internal pull-up resistor. If, on the other hand, the second argument is set to RISING, the function will only be called when the input goes from low to high (when the switch is released).

The event handler function does not stop the main counting loop while it runs; it actually runs in its own separate thread of execution.

Switches often *bounce* when pressed. This means they don't always transition cleanly from open to closed, but bounce between the two, possibly several times, making it appear that the button was pressed multiple times, in very rapid succession, when actually it was pressed only once.

If you keep pressing the button, you'll probably see this reflected in the output as the message appearing more than once for one button press.

The library actually has an option to stop bounce from being a problem, by preventing the interrupt from being triggered again within a certain amount of time. To make use of this feature, just add the extra optional parameter bouncetime to the add_event_de tect call. The value of bouncetime is in milliseconds.

```
GPIO.add_event_detect(18, GPIO.FALLING, callback=my_callback, bouncetime=100)
```

See Also

For more information on using switches with the Raspberry Pi, see Recipe 11.1.

9.13. Controlling GPIO Outputs Using a Web Interface

Problem

You want to control GPIO outputs using a web interface to your Raspberry Pi.

Solution

Use the bottle Python web server library (Recipe 7.16) to create an HTML web interface to control the GPIO port.

To make this recipe, you will need:

- Breadboard and jumper wires (see "Prototyping Equipment" on page 380)
- Three 1kΩ resistors (see "Resistors and Capacitors" on page 380)
- Three LEDs (see "Opto-Electronics" on page 381)
- Tactile push switch (see "Miscellaneous" on page 382)

The breadboard layout for this is shown in Figure 9-17.

To install the bottle library, see Recipe 7.16.

Open an editor (nano or IDLE) and paste in the following code. As with all the program examples in this book, you can also download the program from the Code section of the *Raspberry Pi Cookbook* website (*http://www.raspberrypicookbook.com*), where it is called *web_control.py*. You will also find a program called *web_control_test*. This program simply tests the hardware by flashing the LEDs in sequence and displaying the state of the switch, without any of the complications of the web server.

Figure 9-17. The breadboard layout controlling GPIO outputs from a web page

Here's the *web_control.py* program:

```python
from bottle import route, run
import RPi.GPIO as GPIO

host = '192.168.1.8'

GPIO.setmode(GPIO.BCM)
led_pins = [18, 23, 24]
led_states = [0, 0, 0]
switch_pin = 25

GPIO.setup(led_pins[0], GPIO.OUT)
GPIO.setup(led_pins[1], GPIO.OUT)
GPIO.setup(led_pins[2], GPIO.OUT)
GPIO.setup(switch_pin, GPIO.IN, pull_up_down=GPIO.PUD_UP)

def switch_status():
    state = GPIO.input(switch_pin)
    if state:
        return 'Up'
    else:
        return 'Down'

def html_for_led(led):
    l = str(led)
    result = " <input type='button' onClick='changed(" + l + ")'
    value='LED " + l + "'/>"
    return result

def update_leds():
    for i, value in enumerate(led_states):
```

```
        GPIO.output(led_pins[i], value)

@route('/')
@route('/<led>')
def index(led="n"):
    if led != "n":
        led_num = int(led)
        led_states[led_num] = not led_states[led_num]
        update_leds()
    response = "<script>"
    response += "function changed(led)"
    response += "{"
    response += "  window.location.href='/' + led"
    response += "}"
    response += "</script>"

    response += '<h1>GPIO Control</h1>'
    response += '<h2>Button=' + switch_status() + '</h2>'
    response += '<h2>LEDs</h2>'
    response += html_for_led(0)
    response += html_for_led(1)
    response += html_for_led(2)
    return response

run(host=host, port=80)
```

Before running the program, modify the line near the top that specifies the IP address of your Raspberry Pi:

```
host = '192.168.1.8'
```

You must run the program as superuser:

```
sudo python web_control.py
```

If it starts correctly, you should see a message like this:

```
Bottle server starting up (using WSGIRefServer())...
Listening on http://192.168.1.8:80/
Hit Ctrl-C to quit.
```

Open a browser window from any machine on your network, even the Raspberry Pi itself, and navigate to the IP address of the Raspberry Pi. The web interface shown in Figure 9-18 should appear.

If you click on one of the three LED buttons at the bottom of the screen, you should find that the appropriate LED toggles on and off.

Also, if you hold down the button as you reload the web page, you should see that the text next to *Button* says *Down* rather than *Up*.

Figure 9-18. A web interface to GPIO

Discussion

To understand how the program works, we first need to look at how a web interface works. All web interfaces rely on a server somewhere (in this case, a program on the Raspberry Pi) responding to requests from a web browser.

When the server receives a request, it looks at the information that comes with the request and formulates some HTML (HyperText Markup Language) in response. In the case of this program, the information that the web server program can obtain is:

```
def index(led="n"):
```

If the web request is just to the root page (*http://192.168.1.8/*), then *led* will be given a default value of *n*. However, if we were to browse the URL *http://192.168.1.8/2*, then the *2* on the end of the URL would be assigned to the *led* parameter.

The *led* parameter is then used to determine that LED 2 should be toggled.

To be able to access this LED-toggling URL, we need to arrange things so that when the button for LED 2 is pressed, the page is reloaded with this extra parameter on the end of the URL. The trick here is to include a JavaScript function in HTML that is returned to the browser. When the browser runs this function, it causes the page to be reloaded with the appropriate extra parameter.

This all means that we have the rather mind-bending situation where the Python program is generating code in JavaScript to be run later by the browser. The lines that generate this JavaScript function are as follows:

```
response = "<script>"
response += "function changed(led)"
response += "{"
```

```
response += "  window.location.href='/' + led"
response += "}"
response += "</script>"
```

Rather than repeat the HTML for each of the buttons, this is generated by the function
html_for_led:

```
def html_for_led(led):
    l = str(led)
    result = " <input type='button' onClick='changed(" + l + ")'
    value='LED " + l + "'/>"
    return result
```

This code is used three times, once for each button, and links a button press with the
changed function. The function is also supplied with the LED number as its parameter.

The code that actually changes the GPIO outputs lives in the update_leds function.
This function is called every time the server receives a request that includes an LED
number that requires toggling:

```
def update_leds():
    for i, value in enumerate(led_states):
        GPIO.output(led_pins[i], value)
```

The function simply iterates over an array of states, setting each output in turn to its
current value.

The following line in the index function toggles the value in the state array for the LED
supplied in the led parameter:

```
led_states[led_num] = not led_states[led_num]
```

The process of reporting the state of the push button is much more straightforward. It
is simply a case of reading the state of the input and generating HTML that reports the
button position. This is all contained in the switch_status function:

```
def switch_status():
    state = GPIO.input(switch_pin)
    if state:
        return 'Up'
    else:
        return 'Down'
```

See Also

For more information on using bottle, see Recipe 7.16 and the bottle documenta-
tion (*http://bottlepy.org/docs/dev/*).

Motors

10.0. Introduction

In this chapter, you will investigate the use of many different types of motors with the Raspberry Pi.

10.1. Controlling Servo Motors

 Be sure to check out the accompanying video for this recipe at *http://razzpisampler.oreilly.com*.

Problem

You want to use a Raspberry Pi to control the position of a servo motor.

Solution

Use PWM to control the width of pulses to a servo motor to change its angle. Although this will work, the PWM generated is not completely stable, so there will be a little bit of jitter with the servo.

You should also power the servo from a separate 5V power supply because peaks in the load current are likely to crash or overload the Raspberry Pi.

To make this recipe, you will need:

- 5V servo motor (see "Miscellaneous" on page 382)
- Breadboard and jumper wires (see "Prototyping Equipment" on page 380)
- 1kΩ resistor (see "Resistors and Capacitors" on page 380)

- 5V 1A power supply or 4.8V battery pack (see "Miscellaneous" on page 382)

The breadboard layout for this is shown in Figure 10-1.

Figure 10-1. Controlling a servo motor

The 1kΩ resistor is not essential, but it does protect the GPIO pin from unexpectedly high currents in the control signal, which could occur if a fault developed on the servo.

The leads of the servo may not be the same as the colors indicated in Figure 10-2. It is common for the 5V wire to be red, the ground brown, and the control lead orange.

You can, if you prefer, power the servo from a battery pack rather than a power supply. Using a four-cell AA battery holder with rechargeable batteries will provide around 4.8V and work well with a servo. Using four alkali AA cells to provide 6V will be fine for many servos, but check the datasheet of your servo to make sure it is OK with 6V.

The user interface for setting the angle of the servo is based on the *gui_slider.py* program intended for controlling the brightness of an LED (Recipe 9.2). However, you can modify it so that the slider sets the angle, between 0 and 180 degrees (Figure 10-2).

Figure 10-2. User interface for controlling a servo motor

Open an editor (nano or IDLE) and paste in the following code. As with all the program examples in this book, you can also download the program from the Code section of the *Raspberry Pi Cookbook* website (*http://www.raspberrypicookbook.com*), where it is called *servo.py*.

Note that this program uses a graphical user interface, so you cannot run it from SSH.

You must run it from the windowing environment on the Pi itself or via remote control using VNC (Recipe 2.8). You also need to run it as superuser, so run it with the command `sudo python servo.py`:

```
from Tkinter import *
import RPi.GPIO as GPIO
import time

GPIO.setmode(GPIO.BCM)
GPIO.setup(18, GPIO.OUT)
pwm = GPIO.PWM(18, 100)
pwm.start(5)

class App:

    def __init__(self, master):
        frame = Frame(master)
        frame.pack()
        scale = Scale(frame, from_=0, to=180,
            orient=HORIZONTAL, command=self.update)
        scale.grid(row=0)

    def update(self, angle):
        duty = float(angle) / 10.0 + 2.5
        pwm.ChangeDutyCycle(duty)

root = Tk()
root.wm_title('Servo Control')
app = App(root)
root.geometry("200x50+0+0")
root.mainloop()
```

Discussion

Servo motors are used in remote control vehicles and robotics. Most servo motors are not *continuous*; that is, they cannot rotate all the way around but rather just over an angle of about 180 degrees.

The position of the servo motor is set by the length of a pulse. The servo expects to receive a pulse at least every 20 milliseconds. If that pulse is high for 1 millisecond, the servo angle will be zero; if it is 1.5 milliseconds, it will be at its center position; and if it is 2 milliseconds, it will be at 180 degrees (Figure 10-3).

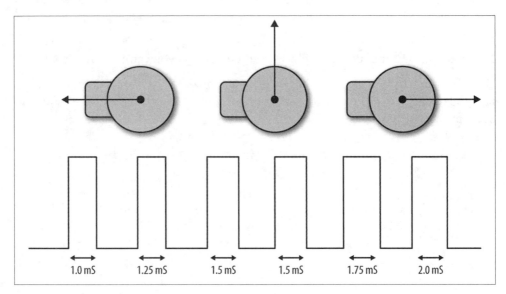

Figure 10-3. Servo motors

The example program sets the PWM frequency to 100 Hz, which will send a pulse to the servo every 10 milliseconds. The angle is converted into a duty cycle between 0 and 100. This actually produces pulses shorter than the 1 millisecond expected minimum value and longer than 2 milliseconds maximum.

See Also

If you have a lot of servos to control, or require greater stability and precision, then you can use a dedicated servo controller module, as described in Recipe 10.2.

Adafruit has developed another method of servo control (*http://bit.ly/17FVspk*).

For more information on using a breadboard and jumper wires with the Raspberry Pi, see Recipe 8.10.

10.2. Controlling a Large Number of Servo Motors

Problem

You want to control a number of servos, and you need high-quality positioning of the servo motor.

Solution

Use a servo controller module like this one from Adafruit: *http://www.adafruit.com/products/815*.

This module allows you to control up to 16 servos or PWM channels using the I2C interface of the Raspberry Pi.

To make this recipe, you will need:

- 1 or more 5V servo motors (see "Miscellaneous" on page 382)
- Adafruit 12-bit PWM/servo driver—I2C interface (see "Modules" on page 381)
- Breadboard and jumper wires (see "Prototyping Equipment" on page 380)
- 5V power supply or 4.8V battery pack (see "Miscellaneous" on page 382)

Figure 10-4 shows how you can wire one of these modules to a Raspberry Pi using a breadboard.

Power is supplied to the logic circuits of the module from the 3.3V connection of the Raspberry Pi. This is entirely separate from the power supply for the servo motors, which comes from an external 5V power adapter.

You can, if you prefer, power the servos from a battery pack rather than a power supply. Using a four-cell AA battery holder with rechargeable batteries will provide around 4.8V and work well with most servos. Using four alkali AA cells to provide 6V will be fine for many servos, but check the datasheet of your servo to make sure it is OK with 6V.

The pin headers for connecting servos are conveniently arranged so that the servo lead will fit directly onto the pins. Be careful to get them the right way around.

To use the Adafruit software for this module, you will need to install Git (Recipe 3.19) and also set up I2C on the Raspberry Pi (Recipe 8.4). Follow both of these recipes before you go any further.

The Adafruit library isn't really a proper library that contains an installation script, but rather just a directory that contains a number of files. So, when using it, you need to be in the directory to which they are downloaded or your program will not find them.

Figure 10-4. Servo motor/PWM module

To download all the Adafruit software library for Raspberry Pi, enter the following:

```
$ git clone https://github.com/adafruit/Adafruit-Raspberry-Pi-Python-Code.git
$ cd Adafruit-Raspberry-Pi-Python-Code
$ cd Adafruit_PWM_Servo_Driver
```

The last two lines set the current directory to the directory containing the code for PWM as well as an example program supplied by Adafruit, which you can run with the command:

```
$ sudo python Servo_Example.py
```

An alternative example next modifies the example program from Recipe 10.2 so that you can use a slider to set the servo position between 0 and 180 degrees. The program file must be saved in the *Adafruit_PWM_Servo_Driver* directory. The slider will change the servo positions of both channels 0 and 1 at the same time, so the two servos should move in the same way when you adjust the slider.

Open an editor (nano or IDLE) and paste in the following code. As with all the program examples in this book, you can also download the program from the Code section of the *Raspberry Pi Cookbook* website (*http://www.raspberrypicookbook.com*), where it is called *servo_module.py*. Note that this program uses a graphical user interface, so you can't run it from SSH. You must run it from the windowing environment on the Pi itself or via remote control using VNC (Recipe 2.8).

```
from Tkinter import *
from Adafruit_PWM_Servo_Driver import PWM
```

```
import time

pwm = PWM(0x40)
pwm.setPWMFreq(50)

class App:

    def __init__(self, master):
        frame = Frame(master)
        frame.pack()
        scale = Scale(frame, from_=0, to=180,
                orient=HORIZONTAL, command=self.update)
        scale.grid(row=0)

    def update(self, angle):
        pulse_len = int(float(angle) * 500.0 / 180.0) + 110
        pwm.setPWM(0, 0, pulse_len)
        pwm.setPWM(1, 0, pulse_len)

root = Tk()
root.wm_title('Servo Control')
app = App(root)
root.geometry("200x50+0+0")
root.mainloop()
```

Discussion

The first line after the import creates a new instance of PWM using the I2C address specified as its argument—in this case, 0×40. The module has solder pad connections that allow you to change the I2C address if this conflicts with another I2C device you are using, or you want to use more than one of these modules.

The next line sets the PWM frequency to 50 Hz, which will provide an update pulse every 20 milliseconds.

The line that actually sets the PWM for a particular channel is:

```
pwm.setPWM(0, 0, pulse_len)
```

The first argument is the PWM channel whose duty cycle is to be changed. Each cycle of PWM is divided into 4,096 ticks, and the second argument is the tick at which the pulse should start. This will always be 0. The third argument is the tick at which the pulse should end. The constants of 500.0 and 110 in the following line were tweaked with a little trial and error to provide a standard servo with as close to 180 degrees of movement as possible:

```
pulse_len = int(float(angle) * 500.0 / 180.0) + 110
```

When selecting a power supply for this module, remember that a standard remote control servo can easily draw 400 mA while it's moving, and more if it's under load. So if

you plan to have a lot of large servos moving at the same time, you will need a big power adapter.

See Also

To control a single servo without using a module, see Recipe 10.1. To control a number of servos at the same time using an Arduino, see Recipe 14.9.

See the Adafruit documentation for its Raspberry Pi library (*http://bit.ly/1iomEey*).

For more information on using a breadboard and jumper wires with the Raspberry Pi, see Recipe 8.10.

10.3. Controlling the Speed of a DC Motor

Problem

You want to control the speed of a DC motor, using your Raspberry Pi.

Solution

You can use the same design as Recipe 9.4. It is, however, a good idea to place a diode across the motor to prevent voltage spikes from damaging the transistor or even the Raspberry Pi (Figure 10-5). The 1N4001 is a suitable diode for this (see "Transistors and Diodes" on page 381). The diode has a stripe at one end, so make sure that this is the right way around.

For low-power motors (less than 200 mA), you can base the design on the one used for a relay in Recipe 9.5. You will need:

- 3V to 12V DC motor
- Breadboard and jumper wires (see "Prototyping Equipment" on page 380)
- 1kΩ resistor (see "Resistors and Capacitors" on page 380)
- Transistor 2N3904 (<see "Transistors and Diodes" on page 381)
- Diode 1N4001 (see "Transistors and Diodes" on page 381)
- Power supply with voltage to match the motor

If you are only using a low-power DC motor (less than 200 mA), you can use a smaller (and cheaper) transistor (Figure 10-6).

You can probably get away with powering a small motor from the 5V supply line on the GPIO connector. If you find that the Raspberry Pi crashes, use an external power supply, as shown in Figure 10-5.

Figure 10-5. Controlling a high-power motor

Figure 10-6. Controlling a low-power motor

To control the speed of the motor, you can use the program from Recipe 9.8 (*gui_slider.py*). You can also download the program from the Code section of *http://www.raspberrypicookbook.com*, where it is called *gui_slider.py*. Note that this program uses a graphical user interface, so you can't run it from SSH. You must run it from the windowing environment on the Pi itself or via remote control using VNC (Recipe 2.8).

Discussion

This is actually the same circuit as you used to switch a relay in Recipe 9.5, except with the motor in place of the relay coil. See Recipe 9.4 for a description of how it works.

See Also

This design only controls the motor's speed. It can't control its direction. For that, you need to see Recipe 10.4.

For more information on using a breadboard and jumper wires with the Raspberry Pi, see Recipe 8.10.

10.4. Controlling the Direction of a DC Motor

Problem

You want to control both the speed and direction of a small DC motor.

Solution

Use an H-Bridge chip or module. See "Discussion" on page 250 for a description of how H-Bridges work.

You have two recipes to choose from for controlling a motor. The first, the "DIY approach," uses solderless breadboard and a L293D chip. The second design uses a ready-made H-Bridge module from SparkFun, connecting it directly to the Raspberry Pi with jumper leads.

Both the L293D and the SparkFun module are capable of driving two motors without any extra hardware.

Option 1: L293D chip and breadboard

If you take the L293D route, you will need:

- 3V to 12V DC motor
- Breadboard and jumper wires (male to female; see "Prototyping Equipment" on page 380)

Figure 10-7. Using a L293D chip to control a motor

- L293D chip (see "Integrated Circuits" on page 381)
- Power supply with voltage to match the motor

The breadboard layout is shown in Figure 10-7.

Make sure the chip goes the right way around: there is a notch in its top, which is the end that should be at the top of the breadboard.

Option 2: Motor controller module

If you decide to use the SparkFun motor controller, or a similar motor controller module, you will need the following:

- 3V to 12V DC motor
- Jumper wires (female to female; see "Prototyping Equipment" on page 380)
- Header pins (see "Miscellaneous" on page 382)
- SparkFun motor driver 1A dual (see "Modules" on page 381)
- Power supply with voltage to match the motor

Figure 10-8. Wiring a SparkFun motor controller

The layout is shown in Figure 10-8. Note that the motor shown is just a DC motor on its own. If the project is driving wheels, then you will typically use a *gearmotor*, which combines a motor and gearbox to reduce the RPM and increase the torque.

The motor controller module comes without header pins attached, so you will need to add these to the board before you get started. The connections are made using female-to-female header leads.

Figure 10-9 shows the module wired up to a small DC gearmotor, as per the wiring diagram of Figure 10-8.

Software

Whichever hardware option you decided upon, you can use the same program to try out the motor. This allows you to enter a letter *f* or *r* and then a single digit between 0 and 9. The motor will then either go forward or backward, at a speed specified by the digit—0 for stopped, 9 for full speed.

```
$ sudo python motor_control.py
Command, f/r 0..9, E.g. f5 :f5
Command, f/r 0..9, E.g. f5 :f1
Command, f/r 0..9, E.g. f5 :f2
Command, f/r 0..9, E.g. f5 :r2
```

Figure 10-9. Using a SparkFun motor controller with a gearmotor

Open an editor (nano or IDLE) and paste in the following code. As with all the program examples in this book, you can also download the program from the Code section of the *Raspberry Pi Cookbook* website (*http://www.raspberrypicookbook.com*), where it is called *motor_control.py*. This program uses the command line, so you can run it from SSH.

If you're using Python 3, change the command raw_input to just input:

```
import RPi.GPIO as GPIO
import time

enable_pin = 18
in1_pin = 23
in2_pin =24

GPIO.setmode(GPIO.BCM)
GPIO.setup(enable_pin, GPIO.OUT)
GPIO.setup(in1_pin, GPIO.OUT)
GPIO.setup(in2_pin, GPIO.OUT)

pwm = GPIO.PWM(enable_pin, 500)
pwm.start(0)

def clockwise():
```

```
        GPIO.output(in1_pin, True)
        GPIO.output(in2_pin, False)

    def counter_clockwise():
        GPIO.output(in1_pin, False)
        GPIO.output(in2_pin, True)

    while True:
        cmd = raw_input("Command, f/r 0..9, E.g. f5 :")
        direction = cmd[0]
        if direction == "f":
            clockwise()
        else:
            counter_clockwise()
        speed = int(cmd[1]) * 10
        pwm.ChangeDutyCycle(speed)
```

Discussion

Before you look at how the program operates, you need to understand a little more about how an H-Bridge works.

Figure 10-10 shows how one works, using switches rather than transistors or a chip. By reversing the polarity across the motor, an H-bridge also reverses the direction in which the motor turns.

Figure 10-10. An H-Bridge

In Figure 10-10, S1 and S4 are closed and S2 and S3 are open. This allows current to flow through the motor, with terminal A being positive and terminal B being negative. If we were to reverse the switches so that S2 and S3 are closed and S1 and S4 are open, then B would be positive and A would be negative, and the motor would turn in the opposite direction.

However, you may have spotted a danger with this circuit. If by some chance S1 and S2 are both closed, then the positive supply will be directly connected to the negative supply and you will have a short-circuit. The same is true if S3 and S4 are both closed at the same time.

Although you can use individual transistors to make an H-Bridge, it is simpler to use an H-Bridge IC such as the L293D. This chip actually has two H-Bridges in it, so you can use it to control two motors.

The L293 has three control pins for each of the two motor control channels. The En able pin just enables or disables the channel as a whole. In the example program, this is connected to a PWM output to control the speed of the motor. The other two pins (IN1 and IN2) control the direction in which the motor will be driven. You can see the use of these two control pins in the functions clockwise and counter_clockwise:

```
def clockwise():
    GPIO.output(in1_pin, True)
    GPIO.output(in2_pin, False)

def counter_clockwise():
    GPIO.output(in1_pin, False)
    GPIO.output(in2_pin, True)
```

If IN1 is high and IN2 is low, the motor will turn in one direction. If those two pins are reversed, the motor will turn in the opposite direction.

See Also

In Recipe 10.8, you see the use of the L293D in a Raspberry Pi interface board, the RaspiRobot Board.

If you just want to control the speed of a single motor, you don't need a whole L293D, and you can just use a single transistor (Recipe 10.3).

Check out the L293D datasheet (*http://bit.ly/18c4GKm*) and the SparkFun Motor Driver Module product page (*http://bit.ly/ILHVkJ*).

For more information on using a breadboard and jumper wires with the Raspberry Pi, see Recipe 8.10.

10.5. Using a Unipolar Stepper Motor

Problem

You want to drive a five-lead unipolar stepper motor using a Raspberry Pi.

Solution

Use a ULN2803 Darlington driver chip.

Stepper motors fit somewhere between DC motors and servo motors in the world of motor technologies. Like a regular DC motor, they can rotate continuously, but you can also very accurately position them by moving them a step at a time in either direction.

To make this recipe, you will need:

- 5V, five-pin unipolar stepper motor (see "Miscellaneous" on page 382)
- ULN2803 Darlington driver IC (see "Integrated Circuits" on page 381)
- Breadboard and jumper wires (see "Prototyping Equipment" on page 380)

Figure 10-11 shows the wiring diagram for using a ULN2803. Note that the chip can be used to drive two such motors. To drive a second stepper motor, you will need to connect four more control pins from the GPIO connector to pins 5 to 8 of the ULN2803 and connect the second motor's four pins to pins 11 to 14 of the ULN2803.

Figure 10-11. Using a ULN2803 to control a unipolar stepper motor

The 5V supply from the GPIO connector may work OK with a small stepper motor. If you experience problems with the Raspberry Pi crashing or need to use a bigger stepper motor, then use a separate supply for the power to the motor (pin 10 of the ULN2803).

Open an editor (nano or IDLE) and paste in the following code. As with all the program examples in this book, you can also download the program from the Code section of the *Raspberry Pi Cookbook* website (*http://www.raspberrypicookbook.com*), where it is called *stepper.py*. This program uses the command line, so you can run it from SSH.

If you're using Python 3, change the command `raw_input` to just `input`:

```
import RPi.GPIO as GPIO
import time

GPIO.setmode(GPIO.BCM)

coil_A_1_pin = 18
coil_A_2_pin = 23
coil_B_1_pin = 24
coil_B_2_pin = 17

GPIO.setup(coil_A_1_pin, GPIO.OUT)
GPIO.setup(coil_A_2_pin, GPIO.OUT)
GPIO.setup(coil_B_1_pin, GPIO.OUT)
GPIO.setup(coil_B_2_pin, GPIO.OUT)

forward_seq = ['1010', '0110', '0101', '1001']
reverse_seq = list(forward_seq) # to copy the list
reverse_seq.reverse()

def forward(delay, steps):
  for i in range(steps):
    for step in forward_seq:
      set_step(step)
      time.sleep(delay)

def backwards(delay, steps):
  for i in range(steps):
    for step in reverse_seq:
      set_step(step)
      time.sleep(delay)

def set_step(step):
  GPIO.output(coil_A_1_pin, step[0] == '1')
  GPIO.output(coil_A_2_pin, step[1] == '1')
  GPIO.output(coil_B_1_pin, step[2] == '1')
  GPIO.output(coil_B_2_pin, step[3] == '1')

while True:
  set_step('0000')
  delay = raw_input("Delay between steps (milliseconds)?")
```

```
steps = raw_input("How many steps forward? ")
forward(int(delay) / 1000.0, int(steps))
set_step('0000')
steps = raw_input("How many steps backwards? ")
backwards(int(delay) / 1000.0, int(steps))
```

When you run the program, you will be prompted for a delay between steps. This should be two or more. You will then be prompted for the number of steps in each direction:

```
$ sudo python stepper.py
Delay between steps (milliseconds)?2
How many steps forward? 100
How many steps backwards? 100
Delay between steps (milliseconds)?10
How many steps forward? 50
How many steps backwards? 50
Delay between steps (milliseconds)?
```

Discussion

Stepper motors use a cogged rotor and electromagnets to nudge the wheel around a *step* at a time (Figure 10-12). Note that the colors of the leads will vary.

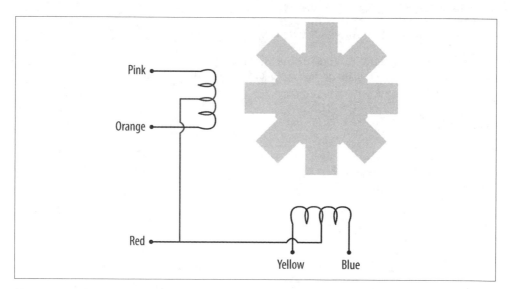

Figure 10-12. A stepper motor

Energizing the coils in a certain order drives the motor around. The number of steps that the stepper motor has in a 360-degree rotation is actually the number of teeth on the rotor.

The example program uses a list of strings to represent each of the four energization stages that make up a single step:

```
forward_seq = ['1010', '0110', '0101', '1001']
```

The sequence for rotating the motor in the opposite direction is just the reverse of the sequence for moving forward.

You can use the forward and backward functions in your programs to step the motor back and forth. The first argument to either function is the delay in milliseconds between each part of the step sequence. This minimum value for this depends on the motor you use. If it's too small, the motor will not turn. Typically, two milliseconds or more will be fine. The second parameter is the number of steps to take.

```
def forward(delay, steps):
    for i in range(steps):
        for step in forward_seq:
            set_step(step)
            time.sleep(delay)
```

The forward function has two nested for loops. The outer one repeats for the number of steps and the inner one iterates over the sequence of motor activations, calling set Step for each in sequence.

```
def set_step(step):
    GPIO.output(coil_A_1_pin, step[0] == '1')
    GPIO.output(coil_A_2_pin, step[1] == '1')
    GPIO.output(coil_B_1_pin, step[2] == '1')
    GPIO.output(coil_B_2_pin, step[3] == '1')
```

The set_step function sets each of the control pins to high or low, depending on the pattern supplied as its argument.

The main loop sets the step to *0000* between moving forward and backward, to set the outputs all to zero when the motor is not actually turning. Otherwise, one of the coils may be left on, causing the motor to draw current unnecessarily.

See Also

If the type of stepper motor you have is a four-wire bipolar stepper motor, see Recipe 10.6.

For more information on stepper motors, the different types, and how they work, see Wikipedia (*http://en.wikipedia.org/wiki/Stepper_motor*), where you will also find a nice animated explanation of the activation pattern for driving the motor.

For information on using servo motors, see Recipe 10.1; for controlling DC motors, see Recipes 10.3 and 10.4.

For more information on using a breadboard and jumper wires with the Raspberry Pi, see Recipe 8.10.

10.6. Using a Bipolar Stepper Motor

Problem

You want to drive a four-lead bipolar stepper motor using a Raspberry Pi.

Solution

Use a L293D H-Bridge driver chip. An H-Bridge is required to drive a bipolar stepper motor because, as the word *bipolar* suggests, the direction of current across the windings needs to be reversed, rather like driving a DC motor in both directions (see Recipe 10.4).

To make this recipe, you will need:

- 12V, four-pin bipolar stepper motor (see "Miscellaneous" on page 382)
- L293D H-Bridge IC (see "Integrated Circuits" on page 381)
- Breadboard and jumper wires (see "Prototyping Equipment" on page 380)

The motor used here, a 12V, is somewhat larger than the previous unipolar stepper motor example. The power for the motor itself is therefore supplied from an external power supply rather than from the Raspberry Pi (see Figure 10-13).

Discussion

You can use the exact same *stepper.py* program to control this stepper (see Recipe 10.5). The design uses both H-Bridges of the L293D, so you need one of these chips for each motor you want to control.

See Also

If the type of stepper motor you have is a five-wire unipolar stepper motor, see Recipe 10.5.

For more information on stepper motors—the different types and how they work—see Wikipedia (*http://en.wikipedia.org/wiki/Stepper_motor*), where you will also find a nice animated explanation of the activation pattern for driving the motor.

For information on using servo motors, see Recipe 10.1; for controlling DC motors, see Recipes 10.3 and 10.4.

You can also drive a stepper motor using the RaspiRobot board (Recipe 10.7).

Figure 10-13. Using a L293D to control a bipolar stepper motor

For more information on using a breadboard and jumper wires with the Raspberry Pi, see Recipe 8.10.

10.7. Using a RaspiRobot Board to Drive a Bipolar Stepper Motor

 Be sure to check out the accompanying video for this recipe at *http://razzpisampler.oreilly.com.*

Problem

You want to control a bipolar stepper motor using a RaspiRobot board.

Solution

The RaspiRobot board uses the same L293D dual H-Bridge chip that you used in Recipe 10.6.

The RaspiRobot board uses the power supply directly from its DC socket as the supply to the motor and regulates that same supply down to 5V to drive the Raspberry Pi. So in this case, the 12V power will be supplying both the 12V stepper motor and the Raspberry Pi.

 The Raspberry Pi should not be powered through its USB connection when the RaspiRobot board is also powered, or slight differences in the 5V from the Raspberry Pi USB and the 5V regulated supply from the RaspiRobot board could cause large currents to flow and damage either the board or the Raspberry Pi. Power one board or the other, but not both.

Connect the stepper motor and power supply to the RaspiRobot board as shown in Figure 10-14. The wire colors for the Adafruit 12V stepper motor are in order, from nearest the DC socket: yellow, red, grey, and green.

Figure 10-14. Using a RaspiRobot board to control a bipolar stepper motor

With a little modification to the pin allocations and step sequence, we can use the program from Recipe 10.6 with a RaspiRobot board.

Open an editor (nano or IDLE) and paste in the following code. As with all the program examples in this book, you can also download the program from the Code section of

the *Raspberry Pi Cookbook* website (*http://www.raspberrypicookbook.com*), where it is called *stepper_rrb.py*. This program uses the command line, so you can run it from SSH.

If you're using Python 3, change the command raw_input to just input:

```python
import RPi.GPIO as GPIO
import time

GPIO.setmode(GPIO.BCM)

coil_A_1_pin = 17
coil_A_2_pin = 4
coil_B_1_pin = 10
coil_B_2_pin = 25

GPIO.setup(coil_A_1_pin, GPIO.OUT)
GPIO.setup(coil_A_2_pin, GPIO.OUT)
GPIO.setup(coil_B_1_pin, GPIO.OUT)
GPIO.setup(coil_B_2_pin, GPIO.OUT)

forward_seq = ['1011', '1111', '1110', '1010']
reverse_seq = list(forward_seq) # to copy the list
reverse_seq.reverse()

def forward(delay, steps):
  for i in range(steps):
    for step in forward_seq:
      set_step(step)
      time.sleep(delay)

def backwards(delay, steps):
  for i in range(steps):
    for step in reverse_seq:
      set_step(step)
      time.sleep(delay)

def set_step(step):
  GPIO.output(coil_A_1_pin, step[0] == '1')
  GPIO.output(coil_A_2_pin, step[1] == '1')
  GPIO.output(coil_B_1_pin, step[2] == '1')
  GPIO.output(coil_B_2_pin, step[3] == '1')

while True:
  set_step('0000')
  delay = raw_input("Delay between steps (milliseconds)?")
  steps = raw_input("How many steps forward? ")
  forward(int(delay) / 1000.0, int(steps))
  set_step('0000')
  steps = raw_input("How many steps backwards? ")
  backwards(int(delay) / 1000.0, int(steps))
```

Discussion

The RaspiRobot board uses the L293D in a different arrangement from how you used it in Recipe 10.4, as it uses a pin to enable each channel (pins 17 and 10), and a second pair of pins that controls the direction of each motor (pins 4 and 25). This means that as well as changing the pin allocations, you also need to modify the step sequence to:

```
forward_seq = ['1011', '1111', '1110', '1010']
```

The first and third bit of each part of the sequence are always 1, enabling both motors. It is now only the second and fourth bits that control the polarity of each of the two stepper windings.

See Also

You can find out more about the RaspiRobot board and other projects that use it at the RaspiRobot website (*http://www.raspirobot.com*).

To drive a stepper motor using a L293D on a breadboard, see Recipe 10.6.

10.8. Building a Simple Robot Rover

Problem

You want to use a Raspberry Pi as the controller for a simple roving robot.

Solution

Use a RaspiRobot board as an interface board to the Raspberry Pi to control two motors and a robot chassis kit such as the Magician Chassis.

To make this recipe, you will need:

- RaspiRobot board (see "Modules" on page 381)
- Magician Chassis with gearmotors (see "Miscellaneous" on page 382)
- Wireless keyboard and mouse

The first step in building the robot is to assemble the Magician Chassis. The chassis is supplied with a four-cell AA battery holder, but to provide power to the Raspberry Pi, you really need a six-cell battery holder, so when you get to that part of the instructions supplied with the Magician Chassis, leave out the battery box.

The RaspiRobot board may be supplied as a kit of parts that must be soldered onto the supplied circuit board. Follow the instructions that come with this board, and then wire everything together as shown in Figure 10-15.

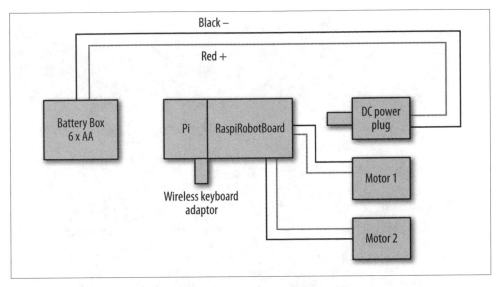

Figure 10-15. Wiring a roving robot

The battery pack will supply power to the RaspiRobot board, which in turn supplies 5V to the Raspberry Pi. So only one power supply is required.

The finished rover should look something like Figure 10-16.

To drive the robot, you are going to use a control program that allows you to steer the rover using the up and down arrow keys on a wireless keyboard. The program for this requires you to download and install the library that accompanies the RaspiRobot board.

The RaspiRobot library requires two other libraries to be installed first, RPi.GPIO (Recipe 8.3) and Pyserial (Recipe 8.8). Enter the following commands into a Terminal session on your Raspberry Pi:

```
$ sudo apt-get install python-rpi.gpio
$ sudo apt-get install python-serial
```

On your Raspberry Pi, issue the following commands in a Terminal window:

```
$ wget https://github.com/simonmonk/raspirobotboard/archive/master.zip
$ unzip master.zip
$ cd raspirobotboard-master
$ sudo python setup.py install
```

Open an editor (nano or IDLE) and paste in the following code. As with all the program examples in this book, you can also download the program from the Code section of the *Raspberry Pi Cookbook* website (*http://www.raspberrypicookbook.com*), where it is called *rover.py*.

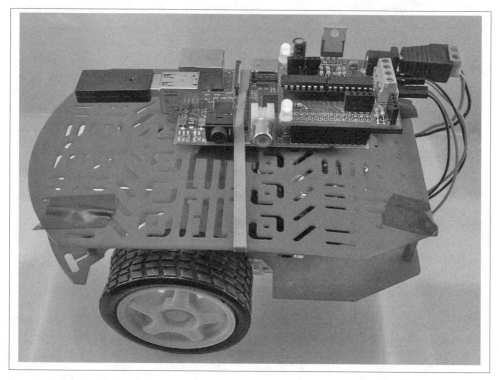

Figure 10-16. The finished robot

To be able to intercept keypresses, this program uses the PyGame library, which does require a user interface (even though the rover won't have a screen attached so, you won't see the interface). For this reason, you can't just launch the program over SSH, but must start it in a VNC session (Recipe 2.8) and/or arrange for the program to autostart on boot (Recipe 3.20):

```
from raspirobotboard import *
import pygame
import sys
from pygame.locals import *

rr = RaspiRobot()

pygame.init()
screen = pygame.display.set_mode((640, 480))

pygame.display.set_caption('RaspiRobot')
pygame.mouse.set_visible(0)

while True:
    for event in pygame.event.get():
```

```
if event.type == QUIT:
    sys.exit()
if event.type == KEYDOWN:
    if event.key == K_UP:
        rr.forward()
        rr.set_led1(True)
        rr.set_led2(True)
    elif event.key == K_DOWN:
        rr.set_led1(True)
        rr.set_led2(True)
        rr.reverse()
    elif event.key == K_RIGHT:
        rr.set_led1(False)
        rr.set_led2(True)
        rr.right()
    elif event.key == K_LEFT:
        rr.set_led1(True)
        rr.set_led2(False)
        rr.left()
    elif event.key == K_SPACE:
        rr.stop()
        rr.set_led1(False)
        rr.set_led2(False)
```

Discussion

You can make the rover more interesting by adding peripherals to it. You could, for example, attach a webcam and set up web streaming so that your roving robot becomes a roving spycam (Recipe 4.5).

There are other ways that you could control your robot, including attaching a WiFi dongle and having the Raspberry Pi serve a web page with buttons on it for left, right, stop, and so on (see Recipe 9.13). You can even drive the robot using a wireless link from a Chronos watch (see GitHub (*http://bit.ly/19zDP55*)).

See Also

You can find out more about the RaspiRobot board and other projects that use it at the RaspiRobot website (*http://www.raspirobot.com*).

Digital Inputs

11.0. Introduction

In this chapter, we look at recipes for using digital inputs, such as switches and keypads. This chapter also covers modules that have a digital output that can be connected to a Raspberry Pi input.

Most of the recipes require the use of a solderless breadboard and male-to-male jumper wires (see Recipe 8.10).

11.1. Connecting a Push Switch

 Be sure to check out the accompanying video for this recipe at *http://razzpisampler.oreilly.com*.

Problem

You want to connect a switch to your Raspberry Pi so that when you press it, some Python code is run.

Solution

Connect a switch to a GPIO pin and use the RPi.GPIO library in your Python program to detect the button press.

To make this recipe, you will need:

- Breadboard and jumper wires (see "Prototyping Equipment" on page 380)
- Tactile push switch (see "Miscellaneous" on page 382)

Figure 11-1 shows how to connect a tactile push switch, using a breadboard and jumper wires.

Figure 11-1. Connecting a push switch to a Raspberry Pi

Open an editor (nano or IDLE) and paste in the following code. As with all the program examples in this book, you can also download the program from the Code section of the *Raspberry Pi Cookbook* website (*http://www.raspberrypicookbook.com*), where it is called *switch.py*.

This example code displays a message when the button is pressed:

```
import RPi.GPIO as GPIO
import time

GPIO.setmode(GPIO.BCM)

GPIO.setup(18, GPIO.IN, pull_up_down=GPIO.PUD_UP)

while True:
    input_state = GPIO.input(18)
    if input_state == False:
```

```
        print('Button Pressed')
        time.sleep(0.2)
```

You will need to run the program as superuser:

```
pi@raspberrypi ~ $ sudo python switch.py
Button Pressed
Button Pressed
Button Pressed
Button Pressed
```

Discussion

You will notice that the switch is wired so that when it is pressed, it will connect pin 18 configured as an input to GND. The input pin is normally pulled up to 3.3V by the optional argument `pull_up_down=GPIO.PUD_UP` in `GPIO.setup`. This means that when you read the input value using `GPIO.input`, `False` will be returned if the button is pressed. This is a little counterintuitive.

Each GPIO pin has software configurable pull-up and pull-down resistors. When using a GPIO pin as an input, you can configure these resistors so that one or either or neither of the resistors is enabled, using the optional `pull_up_down` parameter to `GPIO.setup`. If this parameter is omitted, then neither resistor will be enabled. This leaves the input *floating*, which means that its value cannot be relied upon and it will drift between high and low depending on what it picks up in the way of electrical noise.

If it is set to `GPIO.PUD_UP`, the pull-up resistor is enabled; if it is set to `GPIO.PUD_DOWN`, the pull-down resistor is enabled.

You might expect the push switch to have just two connections, which are either open or closed. While some of these tactile push switches do have just two connections, most have four. Figure 11-2 shows how these connections are arranged.

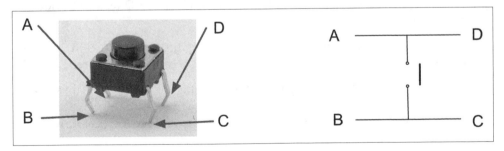

Figure 11-2. A tactile push switch

Actually, there are only really two electrical connections, because inside the switch package pins B and C are connected together, as are A and D.

See Also

For more information on using a breadboard and jumper wires with the Raspberry Pi, see Recipe 8.10.

To use a switch to trigger an interrupt, see Recipe 9.12.

To debounce a switch, see Recipe 11.5.

To use external pull-up or pull-down resistors, see Recipe 11.6.

11.2. Toggling with a Push Switch

Problem

You want to turn something on and off with a push switch so that it toggles between on and off each time you press it.

Solution

Record the last *state* of the button and invert that value each time the button is pressed.

The following example toggles an LED on and off as you press the switch.

To make this recipe, you will need:

- Breadboard and jumper wires (see "Prototyping Equipment" on page 380)
- Tactile push switch (see "Miscellaneous" on page 382)
- LED (see "Opto-Electronics" on page 381)
- 470Ω resistor (see "Resistors and Capacitors" on page 380)

Figure 11-3 shows how to connect a tactile push switch and LED using a breadboard and jumper wires.

As well as the male-to-female jumper wires connecting the Raspberry Pi to the breadboard, you will also need one male-to-male jumper wire or solid core wire.

Open an editor (nano or IDLE) and paste in the following code. As with all the program examples in this book, you can also download the program from the Code section of the *Raspberry Pi Cookbook* website (*http://www.raspberrypicookbook.com*), where it's called *switch_on_off.py*:

```
import RPi.GPIO as GPIO
import time

GPIO.setmode(GPIO.BCM)

switch_pin = 18
```

Figure 11-3. Connecting a push switch and LED to a Raspberry Pi

```
led_pin = 23

GPIO.setup(switch_pin, GPIO.IN, pull_up_down=GPIO.PUD_UP)
GPIO.setup(led_pin, GPIO.OUT)

led_state = False
old_input_state = True # pulled-up

while True:
    new_input_state = GPIO.input(switch_pin)
    if new_input_state == False and old_input_state == True:
        led_state = not led_state
    old_input_state = new_input_state
    GPIO.output(led_pin, led_state)
```

Discussion

The variable led_state contains the current state of the LED (True for On and False for Off). Whenever the button is pressed, the following line is run:

```
led_state = not led_state
```

The `not` command inverts the value of `led_state`, so if `led_state` is `True`, it becomes `False` and vice versa.

The variable `old_input_state` is used to remember the button position so that a button press is defined as occurring only when the input state changes from being `True` (switch not pressed) to `False` (switch pressed).

See Also

You will find that sometimes when you press the button, it does not seem to toggle the LED. This is because of switch bounce. You can find other techniques for avoiding switch bounce in Recipe 11.5.

11.3. Using a Two-Position Toggle or Slide Switch

Problem

You want to connect a two-position toggle or slide switch to your Raspberry Pi, and be able to find the position of the switch in your Python program.

Solution

Use the switch as you would a tactile push switch (Recipe 11.1), just connecting the center and one end contact (Figure 11-4).

To make this recipe, you will need:

- Breadboard and jumper wires (see "Prototyping Equipment" on page 380)
- Miniature toggle or slide switch (see "Miscellaneous" on page 382)

The same code we used in Recipe 11.1 works with this arrangement.

Discussion

These type of slide switches are useful, because you can see the position they are set to without the need for some additional indicator like an LED. However, they are more fragile and a little more expensive than the tactile push switches, which are used more and more in consumer electronics because they can sit behind a nicer-looking plastic button.

See Also

To use a three-position switch, with a center-off position, see Recipe 11.4.

Figure 11-4. Connecting a slide switch to a Raspberry Pi

11.4. Using a Center-Off Toggle or Slide Switch

Problem

You want to connect a three-position (center-off) toggle switch to your Raspberry Pi, and be able to find the position of the switch in your Python program.

Solution

Connect the switch to two GPIO pins, as shown in Figure 11-5, and use the RPi.GPIO library in your Python program to detect the position of the switch.

To make this recipe, you will need:

- Breadboard and jumper wires (see "Prototyping Equipment" on page 380)
- Miniature center-off three-position toggle switch (see "Miscellaneous" on page 382)

The common (center) connection of the switch is connected to ground, and each of the two ends of the switch are connected to a GPIO pin with the internal pull-up resistor enabled.

Open an editor (nano or IDLE) and paste in the following code. As with all the program examples in this book, you can also download the program from the Code section of the *Raspberry Pi Cookbook* website (*http://www.raspberrypicookbook.com*), where it is called *switch_2_pos.py*:

Figure 11-5. Connecting a three-position switch to a Raspberry Pi

```
import RPi.GPIO as GPIO
import time

GPIO.setmode(GPIO.BCM)

top_input = 18
bottom_input = 23

GPIO.setup(top_input, GPIO.IN, pull_up_down=GPIO.PUD_UP)
GPIO.setup(bottom_input, GPIO.IN, pull_up_down=GPIO.PUD_UP)

switch_position = "unknown"

while True:
    top_state = GPIO.input(top_input)
    bottom_state = GPIO.input(bottom_input)
    new_switch_position = "unknown"
    if top_state == False:
        new_switch_position = "up"
    elif bottom_state == False:
        new_switch_position = "down"
    else:
        new_switch_position = "center"
    if new_switch_position != switch_position:
        switch_position = new_switch_position
        print(switch_position)
```

Run the program, and as you move the switch from top to center to bottom, the position of the switch will be reported every time it changes:

```
$ sudo python switch_3_pos.py
up
```

```
center
down
```

Discussion

The program sets up two inputs with pull-up resistors enabled. The variable `switch_po` `sition` is used to record the current position of the switch.

Inside the loop, both GPIO inputs are read and the three conditions of the `if`, `elif`, and `else` structure determine the position of the switch, assigning the value to a variable called `new_switch_position`. If this differs from the previous value, then the switch position is printed.

You will find a wide range of types of toggle switches. Some will be described as DPDT, SPDT, SPST, or SPST, momentary on, and so on. The meaning of these letters is as follows:

- D—Double
- S—Single
- P—Pole
- T—Throw

A DPDT switch is double pole, double throw. The word *pole* refers to the number of separate switch contacts that are controlled from the one mechanical lever. So, a double pole switch can switch two things on and off independently. A single throw switch can only open or close a single contact (or two contacts if it is double pole). However, a double throw switch can connect the common contact to one of two other contacts.

Figure 11-6 shows the most common types of switch.

See Also

For more information on how `if` statements work, see Recipe 5.18. For the most basic switch recipe, see Recipe 11.1.

11.5. Debouncing a Button Press

Problem

Sometimes when you press the button on a switch, the expected action happens more than once, because the switch contacts *bounce*. In that case, you want to write code to *de-bounce* the switch.

Figure 11-6. Types of toggle switches

Solution

There are a number of solutions to this problem. To explore them, build the breadboard setup from Recipe 11.2.

The original code for this example, without any attempt at debouncing, looks like this:

```
import RPi.GPIO as GPIO
import time

GPIO.setmode(GPIO.BCM)

switch_pin = 18
led_pin = 23

GPIO.setup(switch_pin, GPIO.IN, pull_up_down=GPIO.PUD_UP)
GPIO.setup(led_pin, GPIO.OUT)

led_state = False
old_input_state = True # pulled-up
```

```
while True:
    new_input_state = GPIO.input(switch_pin)
    if new_input_state == False and old_input_state == True:
        led_state = not led_state
    old_input_state = new_input_state
    GPIO.output(led_pin, led_state)
```

The problem is that if the switch contacts bounce, it is just as if the switch were pressed more than once in very rapid succession. If they bounce an odd number of times, then things will seem to be OK. But if they bounce an even number of times, the two events will toggle the LED on and then straight back off again.

You need to ignore any changes after the switch is pressed for a short amount of time, while the switch finishes bouncing.

The quick and easy way to do this is to introduce a short sleep after the button press is detected by adding a time.sleep command of, say, 0.2 seconds. This delay is probably much higher than necessary, strictly speaking. You may find that you can reduce this considerably.

```
import RPi.GPIO as GPIO
import time

GPIO.setmode(GPIO.BCM)

switch_pin = 18
led_pin = 23

GPIO.setup(switch_pin, GPIO.IN, pull_up_down=GPIO.PUD_UP)
GPIO.setup(led_pin, GPIO.OUT)

led_state = False
old_input_state = True # pulled-up

while True:
    new_input_state = GPIO.input(switch_pin)
    if new_input_state == False and old_input_state == True:
        led_state = not led_state
        time.sleep(0.2)
    old_input_state = new_input_state
    GPIO.output(led_pin, led_state)
```

Discussion

This solution is OK in most situations, but you might also consider using the switch pin with an interrupt (Recipe 9.12).

Switch bouncing occurs on most switches and can be quite severe on some switches, as the oscilloscope trace in Figure 11-7 shows.

Figure 11-7. Contact bounce with a poor switch

You can see that there is contact bounce both as the switch closes and when it is released. Most switches are not as bad as this one.

See Also

For the basics of connecting a button, see Recipe 11.1.

11.6. Using an External Pull-up Resistor

Problem

You want to run a long wire from the Raspberry Pi to the switch, but you are getting some false readings on the input pin.

Solution

The internal pull-up resistors are quite weak (about 40kΩ). So, if you run a long lead to the switch or operate in an electrically noisy environment, you may get false triggerings on the digital input. You can overcome this by turning off the internal pull-up and pull-down resistors and using an external pull-up resistor.

Figure 11-8 shows the use of an external pull-up resistor.

Figure 11-8. Using an external pull-up resistor

To test out this hardware, you can use the program *switch.py*; see Recipe 11.1.

Discussion

The lower the resistance of the resistor, the longer the range of your switch. However, when you press the button, a current flows from 3.3V through the resistor to ground. A 100Ω resistor will draw a current of 3.3V/100Ω = 33 mA. This is within the safe limit for the 3.3V supply of 50 mA, so don't use a lower value than this.

In almost all cases, a 1kΩ resistor will provide a long range with no problems.

See Also

For the basics of connecting a button, see Recipe 11.1.

11.7. Using a Rotary (Quadrature) Encoder

Problem

You want to detect rotation using a rotary encoder.

Solution

Use a rotary (quadrature encoder) connected to two GPIO pins, as shown in Figure 11-9.

Figure 11-9. Connecting a rotary encoder

To make this recipe, you will need:

- Breadboard and jumper wires (see "Prototyping Equipment" on page 380)
- Rotary encoder (quadrature type; see "Miscellaneous" on page 382)

This type of rotary encoder is called a *quadrature encoder*, and it behaves like a pair of switches. The sequence in which they open and close as the rotary encoder's shaft is turned determines the direction of rotation.

The rotary encoder shown has the center lead as the *common* lead and the two leads on either side as A and B. Not all rotary encoders use this layout, so check the pinout on the datasheet for the rotary encoder that you are using. The issue is often confused further because many rotary encoders include a push switch, which will have a separate pair of contacts.

Open an editor (nano or IDLE) and paste in the following code. As with all the program examples in this book, you can also download the program from the Code section of the *Raspberry Pi Cookbook* website (*http://www.raspberrypicookbook.com*), where it is called *rotary_encoder.py*.

```
import RPi.GPIO as GPIO
import time

GPIO.setmode(GPIO.BCM)

input_A = 18
input_B = 23

GPIO.setup(input_A, GPIO.IN, pull_up_down=GPIO.PUD_UP)
GPIO.setup(input_B, GPIO.IN, pull_up_down=GPIO.PUD_UP)
```

```
old_a = True
old_b = True

def get_encoder_turn():
    # return -1, 0, or +1
    global old_a, old_b
    result = 0
    new_a = GPIO.input(input_A)
    new_b = GPIO.input(input_B)
    if new_a != old_a or new_b != old_b :
        if old_a == 0 and new_a == 1 :
            result = (old_b * 2 - 1)
        elif old_b == 0 and new_b == 1 :
            result = -(old_a * 2 - 1)
    old_a, old_b = new_a, new_b
    time.sleep(0.001)
    return result

x = 0

while True:
    change = get_encoder_turn()
    if change != 0 :
        x = x + change
            print(x)
```

The test program simply counts up as you turn the rotary encoder clockwise and counts down when you rotate it counterclockwise.

```
pi@raspberrypi ~ $ sudo python rotary_encoder.py
1
2
3
4
5
6
7
8
9
10
9
8
7
6
5
4
```

Figure 11-10. How quadrature encoders work

Discussion

Figure 11-10 shows the sequence of pulses that you will get from the two contacts, A and B. You can see that the pattern repeats itself after four steps (hence the name *quadrature* encoder).

When rotating clockwise (left to right in Figure 11-10), the sequence will be:

Phase	A	B
1	0	0
2	0	1
3	1	1
4	1	0

When rotating in the opposite direction, the sequence of phases will be reversed.

Phase	A	B
4	1	0
3	1	1
2	0	1
1	0	0

The Python program listed previously implements the algorithm for determining the rotation direction in the function get_encoder_turn. The function will return 0 (if there has been no movement), 1 for a rotation clockwise, or -1 for a rotation counterclockwise. It uses two global variables, old_a and old_b, to store the previous states of the switches A and B; by comparing them with the newly read values, it can determine (using a bit of clever logic) which direction the encoder is turning.

The sleep period of 1 millisecond is to ensure that the next new sample does not occur too soon after the previous sample; otherwise, the transitions can give false readings.

The test program should work reliably no matter how fast you twiddle the knob on the rotary encoder; however, try to avoid doing anything time-consuming in the loop, or you may find that turn steps are missed.

See Also

You can also measure the rotated position of a knob by using a variable resistor with the step response method (Recipe 12.1) or by using an analog-to-digital converter (Recipe 12.4).

11.8. Using a Keypad

Problem

You want to interface a keypad with your Raspberry Pi.

Solution

Keypads are arranged in rows and columns, with a push switch on the intersection of each row or column. To find out which key is pressed, you first connect all the row and column connections to Raspberry Pi GPIO pins. So, for a 4 × 3 keypad, you will need four + three pins. By scanning each column in turn (setting it to output high) and reading the value of each of the row inputs, you can determine which (if any) key is pressed.

Note that keypads show considerable variation in their pinouts.

To make this recipe, you will need:

- Breadboard and jumper wires (see "Prototyping Equipment" on page 380)
- 4 × 3 keypad (see "Miscellaneous" on page 382)
- Seven male header pins (see "Miscellaneous" on page 382)

Figure 11-11 shows the wiring diagram for the project using the SparkFun keypad listed in "Miscellaneous" on page 382. The keypad is supplied without header pins, which must be soldered onto the keypad.

Open an editor (nano or IDLE) and paste in the following code. As with all the program examples in this book, you can also download the program from the Code section of the *Raspberry Pi Cookbook* website (*http://www.raspberrypicookbook.com*), where it is called *keypad.py*.

 Before you run the program, make sure that the row and column pins are correct for the keypad that you are using, and if necessary change the values in the variables rows and cols. If you do not do this, then it is possible that pressing a key could short one GPIO output to another, where one is high and the other is low. This would likely damage your Raspberry Pi.

Figure 11-11. Keypad wiring diagram

Figure 11-12. Keypad pin connections

The row and columns defined here are correct for the SparkFun keypad listed in Appendix A. The first row is connected to GPIO pin 17, the second to 25, and so on. The wiring of the row and column to the keypad connector is illustrated in Figure 11-12.

```
import RPi.GPIO as GPIO
import time

GPIO.setmode(GPIO.BCM)

rows = [17, 25, 24, 23]
cols = [27, 18, 22]
keys = [
    ['1', '2', '3'],
    ['4', '5', '6'],
    ['7', '8', '9'],
    ['*', '0', '#']]

for row_pin in rows:
    GPIO.setup(row_pin, GPIO.IN, pull_up_down=GPIO.PUD_DOWN)

for col_pin in cols:
    GPIO.setup(col_pin, GPIO.OUT)

def get_key():
    key = 0
    for col_num, col_pin in enumerate(cols):
        GPIO.output(col_pin, 1)
        for row_num, row_pin in enumerate(rows):
            if GPIO.input(row_pin):
                key = keys[row_num][col_num]
        GPIO.output(col_pin, 0)
    return key

while True:
    key = get_key()
    if key :
        print(key)
    time.sleep(0.3)
```

You must run the program with superuser privileges, as it accesses the GPIO. You can see the trace from the program pressing each key in turn.

```
pi@raspberrypi ~ $ sudo python keypad.py
1
2
3
4
5
6
7
8
9
*
0
#
```

Discussion

The `keys` variable contains a map of the key name for each row and column position. You can customize this for your keypad.

Because there are quite a lot of pins to initialize as inputs or outputs, both the row and column pins are initialized in loops.

All the real action takes place in the `get_key` function. This enables each column in turn, by setting it to high. An inner loop then tests each of the rows in turn. If one of the rows is high, then the key name corresponding to that row and column is looked up in the `keys` array. If no keypress is detected, then the default value of `key` (0) is returned.

The main `while` loop just gets the key value and prints it. The `sleep` command just slows down the output.

See Also

An alternative to adding a keyboard is simply to use a USB keypad.

11.9. Detecting Movement

Problem

You want to trigger some action in Python when movement is detected.

Solution

Use a PIR (passive infrared) motion detector module.

To make this recipe, you will need:

- Female-to-female jumper wires (see "Prototyping Equipment" on page 380)
- PIR motion detector module (see "Modules" on page 381)

Figure 11-13 shows how the sensor module is wired. This module expects a power supply of 5V and has an output of 3.3V, making it ideal for use with a Raspberry Pi.

 Make sure that the PIR module you use has a 3.3V output. If it has 5V output, you will need to use a pair of resistors to reduce this to 3.3V (see Recipe 8.12).

Figure 11-13. Wiring a PIR motion detector

Open an editor (nano or IDLE) and paste in the following code. As with all the program examples in this book, you can also download the program from the Code section of the *Raspberry Pi Cookbook* website (*http://www.raspberrypicookbook.com*), where it is called *pir.py*.

```
import RPi.GPIO as GPIO
import time

GPIO.setmode(GPIO.BCM)

GPIO.setup(18, GPIO.IN)

while True:
    input_state = GPIO.input(18)
    if input_state == True:
        print('Motion Detected')
        time.sleep(1)
```

The program simply prints out the state of the GPIO input 18.

```
$ sudo python pir.py
Motion Detected
Motion Detected
```

Discussion

Once triggered, the output of the PIR sensor will stay high for a little while. You can adjust this using one of the trimpots on its circuit board.

See Also

You could combine this recipe with Recipe 7.15 to send an email when an intruder is detected.

11.10. Adding GPS to the Raspberry Pi

Problem

You want to connect a serial GPS module to a Raspberry Pi and access the data using Python.

Solution

A 3.3V serial GPS module can be connected directly to Raspberry Pi's RXD connection. However, to be able to use the serial port, you need to disable the console logging feature, so first follow Recipe 8.7.

Figure 11-14 shows how the module is wired. The RXD of the Raspberry Pi connects to Tx of the GPS module. The only other connections are for GND and 5V, so we can easily just use three female-to-female headers.

Figure 11-14. Wiring a GPS to a Raspberry Pi

GPS messages require some decoding. Fortunately, there is a good suite of tools to help us do this. Install the following packages:

```
$ sudo apt-get install gpsd
$ sudo apt-get install gpsd-clients
```

The most important of these is *gpsd*. This is a tool that reads GPS data from a serial or USB connection as well as other sources and makes it available for client programs to use by providing a local web service on port 2748.

Figure 11-15. Testing GPS with cgps

To start the *gpsd* service running, issue the following command:

```
$ sudo gpsd /dev/ttyAMA0
```

You can see if it is working by entering the command:

```
$ cgps -s
```

The `-s` is optional; it just suppresses the display of the raw data (see Figure 11-15).

The third package that we installed (*python-gps*) is, as you might expect, a Python library for accessing the GPS data in a nice, convenient form. We will use *python_gps* with a short test program to display just the latitude, longitude, and time.

Open an editor (nano or IDLE) and paste in the following code. Do not call the file *gps.py*, or it will conflict with the Python GPS library. As with all the program examples in this book, you can also download the program from the Code section of the *Raspberry Pi Cookbook* website (*http://www.raspberrypicookbook.com*), where it is called *gps_test.py*.

```
from gps import *
session = gps()
session.stream(WATCH_ENABLE|WATCH_NEWSTYLE)

while True:
    report = session.next()
```

```
      if report.keys()[0] == 'epx' :
          lat = float(report['lat'])
          lon = float(report['lon'])
          print("lat=%f\tlon=%f\ttime=%s" % (lat, lon, report['time']))
          time.sleep(0.5)
```

Run the program, and you should see some trace like this:

```
$ python gps_test.py
lat=53.710257    lon=-2.664245   time=2013-08-01T08:06:24.960Z
lat=53.710258    lon=-2.664252   time=2013-08-01T08:06:25.960Z
lat=53.710258    lon=-2.664252   time=2013-08-01T08:06:25.960Z
lat=53.710248    lon=-2.664243   time=2013-08-01T08:06:26.960Z
lat=53.710248    lon=-2.664243   time=2013-08-01T08:06:26.960Z
lat=53.710248    lon=-2.664250   time=2013-08-01T08:06:27.960Z
```

Discussion

The program creates a *session* and then establishes a stream of data to be read. The GPS will repeatedly spit out messages in different formats. The if command selects just the messages we want, those that contain the positional information. The parts of the message are stored in a dictionary from which the fields can be accessed and displayed.

Besides using the GPS data with Python, you can also use the xgps tool to display the GPS data (Figure 11-16). Just enter the command:

```
$ xgps
```

This utility requires a display, so you should run it from the Raspberry Pi itself or using VNC (Recipe 2.8).

See Also

You can use the same approach using a USB GPS module (*http://bit.ly/1edSyHX*).

Find out more about *gpsd* (*http://bit.ly/1fP61q4*).

11.11. Intercepting Keypresses

Problem

You want to intercept individual keypresses on a USB keyboard or numeric keypad.

Solution

There are at least two ways to solve this problem. The most straightforward is to use the sys.stdin.read function. This has the advantage over the other method of not requiring a graphical user interface to be running, so a program using it can be run from an *ssh* session.

Figure 11-16. Viewing GPS with xgps

Open an editor (nano or IDLE) and paste in the following code. As with all the program examples in this book, you can also download the program from the Code section of the *Raspberry Pi Cookbook* website (*http://www.raspberrypicookbook.com*), where it is called *keys_sys.py*.

```
import sys, tty, termios

def read_ch():
    fd = sys.stdin.fileno()
    old_settings = termios.tcgetattr(fd)
    try:
        tty.setraw(sys.stdin.fileno())
```

```
        ch = sys.stdin.read(1)
    finally:
        termios.tcsetattr(fd, termios.TCSADRAIN, old_settings)
    return ch

while True:
    ch = read_ch()
    if ch == 'x':
        break
    print("key is: " + ch)
```

The alternative to this is to use Pygame. Pygame is a Python library intended for writing games. It can also be used to detect keypresses. You could then use this to perform some action (or steer a robot, as it's used in Recipe 10.8).

The following example program illustrates the use of Pygame to print out a message each time a key is pressed. However, it only works if the program has access to the windowing system, so you will need to run it using VNC (Recipe 2.8) or directly on the Raspberry Pi. The example code is in the file *keys_pygame*.

```
import pygame
import sys
from pygame.locals import *

pygame.init()
screen = pygame.display.set_mode((640, 480))
pygame.mouse.set_visible(0)

while True:
    for event in pygame.event.get():
        if event.type == QUIT:
            sys.exit()
        if event.type == KEYDOWN:
            print("Code: " + str(event.key) + " Char: " + chr(event.key))
```

This opens a blank Pygame window, and keys will only be intercepted if this has focus. The program produces output in the Terminal window from which the program is run.

If you press an arrow key or Shift key, the program will throw an error because those keys don't have an ASCII value.

```
$python keys_pygame.py
Code: 97 Char: a
Code: 98 Char: b
Code: 99 Char: c
Code: 120 Char: x
Code: 13 Char:
```

In this case, Ctrl-C won't stop this program from running. To stop the program, click the *X* on the PyGame window.

Discussion

When you are using the Pygame approach, other keys have constant values defined for them, which allows you to use the cursor and other non-ASCII keys (like the up arrow key and Home) on the keyboard. This isn't possible with the other approach.

See Also

Intercepting keyboard events can also be an alternative to using a matrix keypad (Recipe 11.8).

11.12. Intercepting Mouse Movements

Problem

You want to detect mouse movements in Python.

Solution

The solution to this is very similar to that of using Pygame to intercept keyboard events (Recipe 11.11).

Open an editor (nano or IDLE) and paste in the following code. As with all the program examples in this book, you can also download the program from the Code section of the *Raspberry Pi Cookbook* website (*http://www.raspberrypicookbook.com*), where it is called *mouse_pygame.py*.

```
import pygame
import sys
from pygame.locals import *

pygame.init()
screen = pygame.display.set_mode((640, 480))
pygame.mouse.set_visible(0)

while True:
    for event in pygame.event.get():
        if event.type == QUIT:
            sys.exit()
        if event.type == MOUSEMOTION:
            print("Mouse: (%d, %d)" % event.pos)
```

The MOUSEMOTION event is triggered whenever the mouse moves within the Pygame window. You can find the coordinates from the pos value of the event. The coordinates are absolute coordinates relative to the top-left corner of the window:

```
Mouse: (262, 285)
Mouse: (262, 283)
```

```
Mouse: (262, 281)
Mouse: (262, 280)
Mouse: (262, 278)
Mouse: (262, 274)
Mouse: (262, 270)
Mouse: (260, 261)
Mouse: (258, 252)
Mouse: (256, 241)
Mouse: (254, 232)
```

Discussion

Other events that you can intercept are MOUSEBUTTONDOWN and MOUSEBUTTONUP. These can be used to detect when the left mouse button has been pressed or released.

See Also

The Pygame documentation for mouse can be found at the Pygame website (*http://www.pygame.org/docs/ref/mouse.html*).

For information on intercepting keyboard events using Pygame, see Recipe 11.11.

11.13. Using a Real-Time Clock Module

Problem

You want your Raspberry Pi to remember the time, even when it has no network connection.

Solution

Use an RTC (real-time clock) module.

A very common RTC chip is the DS1307. It has an I2C interface and can be bought as a ready-to-use module that includes the chip itself, a quartz crystal for it to maintain accurate timekeeping, and a battery holder for a 3V lithium cell.

To make this recipe, you will need:

- A DS1307 or compatible RTC module (see "Modules" on page 381). You can also use an aLaMode board (Recipe 14.13).
- Female-to-female jumper wires (see "Prototyping Equipment" on page 380)

 The RTC module that you are using must be 3.3V compatible. That means that its I2C interface should either have no pull-up resistors at all, or they should pull up to 3.3V and NOT 5V. When using the Adafruit model here, simply don't include the two resistors when soldering together the module. If you have a ready-made module, carefully remove any pull-up resistors.

Assemble the RTC module if it is in kit form, remembering to omit pull-up resistors and then connect the module to the Raspberry Pi, as shown in Figure 11-17.

Figure 11-17. Connecting an RTC module

The DS1307 is an I2C module, so your Raspberry Pi must be set up to work with I2C (see Recipe 8.4). You can check that the device is visible using I2C tools (Recipe 8.5).

```
$ sudo i2cdetect -y 1
     0  1  2  3  4  5  6  7  8  9  a  b  c  d  e  f
00:          -- -- -- -- -- -- -- -- -- -- -- -- --
10: -- -- -- -- -- -- -- -- -- -- -- -- -- -- -- --
20: -- -- -- -- -- -- -- -- -- -- -- -- -- -- -- --
30: -- -- -- -- -- -- -- -- -- -- -- -- -- -- -- --
40: -- -- -- -- -- -- -- -- -- -- -- -- -- -- -- --
50: -- -- -- -- -- -- -- -- -- -- -- -- -- -- -- --
60: -- -- -- -- -- -- -- -- 68 -- -- -- -- -- -- --
70: -- -- -- -- -- -- -- --
```

The *68* in the table indicates that the RTC module is connected to the I2C bus at hex address 68.

If you are using one of the original Raspbery Pi model B, revision 1, boards, then use 0 after the y option in the preceding line. The revision 1 boards were distinctive for having a black audio socket.

You now need to run the following commands so that the RTC can be used with a program called *hwclock*.

```
$ sudo modprobe rtc-ds1307
$ sudo bash
$ echo ds1307 0x68 > /sys/class/i2c-adapter/i2c-1/new_device
```

Once again, if you are using a revision 1 board, change i2c-1 to i2c-0.

You can now access the RTC using the following command:

```
$ sudo hwclock -r
Sat 01 Jan 2000 00:08:08 UTC  -0.293998 seconds
```

As you can see, the clock is not currently set.

To set the time on the RTC module, you first need to make sure that your Raspberry Pi has the right time. If your Pi is connected to the Internet, this should happen automatically. You can check this using the date command:

```
$ date
Tue Aug 20 06:42:47 UTC 2013
```

If the time is wrong, you can also set it manually using date (Recipe 3.33). To transfer the Raspbery Pi system time onto the RTC module, use the following command:

```
$ sudo hwclock -w
```

You can then read the time using the -r option:

```
$ sudo hwclock -r
Wed 02 Jan 2013 03:11:43 UTC  -0.179786 seconds
```

The RTC having the correct time is pointless unless it is used to set the correct system time in Linux when it reboots. To arrange for this to happen, you need to make a few configuration changes.

First, edit */etc/modules* (using sudo nano /etc/modules) and add *rtc-ds1307* to the end of the list of modules. If you have already added some modules while setting up I2C, SPI, and other options, your file might look something like this:

```
# /etc/modules: kernel modules to load at boot time.
#
# This file contains the names of kernel modules that should be loaded
# at boot time, one per line. Lines beginning with "#" are ignored.
# Parameters can be specified after the module name.

snd-bcm2835
i2c-bcm2708
```

```
i2c-dev
spidev
rtc-ds1307
```

Next, you need two commands to run automatically during startup to set the system time. So edit the file */etc/rc.local* using the command sudo nano /etc/rc.local, and just before the final exit 0 line, insert the following two lines. Remember to change i2c-1 to i2c-0 if you are using a revision 1 Pi.

```
$ echo ds1307 0x68 > /sys/class/i2c-adapter/i2c-1/new_device
$ sudo hwclock -s
```

When you finish, the file should look something like this:

```
#
# In order to enable or disable this script just change the execution
# bits.
#
# By default this script does nothing.

# Print the IP address
_IP=$(hostname -I) || true
if [ "$_IP" ]; then
  printf "My IP address is %s\n" "$_IP"
fi
echo ds1307 0x68 > /sys/class/i2c-adapter/i2c-1/new_device
sudo hwclock -s
exit 0
```

Now, when you reboot, your Raspbery Pi should set its system time from the RTC. However, if there is an Internet connection available, this will take precedence for setting the time.

Discussion

An RTC is not essential for a Raspberry Pi by any means, since a Raspberry Pi that is connected to the Internet will automatically connect to a network time server and set its own clock. However, you may not always be using the Pi when connected to a network, in which case a hardware RTC is a good option.

The aLaMode (Recipe 14.13) Arduino-based interface board for Raspberry Pi also includes an RTC module that will work with this arrangement.

See Also

AB Electronics have a neat RTC that plugs directly into the GPIO socket (*http://bit.ly/1hShxWi*). This is shown in Figure 11-18.

Figure 11-18. AB Electronics RTC module

This recipe is based on a tutorial from Adafruit (*http://bit.ly/IcKvQ9*).

Sensors

12.0. Introduction

In this chapter, you will look at recipes for using sensors of various types that will allow the Raspberry Pi to measure temperature, light, and more.

Compared to a board like the Arduino, the Raspberry Pi lacks analog inputs. This means that for many sensors, it is necessary to use additional analog-to-digital (ADC) converter hardware. Fortunately, this is relatively easy to do. It is also possible to use resistive sensors with a capacitor and a couple of resistors.

Most of the recipes will require the use of a solderless breadboard and male-to-female jumper wires (see Recipe 8.10).

12.1. Using Resistive Sensors

 Be sure to check out the accompanying video for this recipe at *http://razzpisampler.oreilly.com*.

Problem

You want to connect a variable resistor to a Raspberry Pi and measure the position of its rotation.

Solution

You can measure resistance on a Raspberry Pi using nothing more than a capacitor, a couple of resistors, and two GPIO pins. In this case, you will be able to read the position of the knob on a small variable resistor (trimpot).

To make this recipe, you will need:

- Breadboard and jumper wires (see "Prototyping Equipment" on page 380)
- 10kΩ trimpot (see "Resistors and Capacitors" on page 380)
- Two 1kΩ resistors (see "Resistors and Capacitors" on page 380)
- 220 nF capacitor (see "Resistors and Capacitors" on page 380)

Figure 12-1 shows the arrangement of components on the breadboard.

Figure 12-1. Measuring resistance on a Raspberry Pi

Open an editor (nano or IDLE) and paste in the following code. As with all the program examples in this book, you can also download the program from the Code section of the *Raspberry Pi Cookbook* website (*http://www.raspberrypicookbook.com*), where it is called *pot_step.py*.

```
import RPi.GPIO as GPIO
import time

GPIO.setmode(GPIO.BCM)

a_pin = 18
b_pin = 23

def discharge():
    GPIO.setup(a_pin, GPIO.IN)
    GPIO.setup(b_pin, GPIO.OUT)
    GPIO.output(b_pin, False)
    time.sleep(0.005)

def charge_time():
```

```
        GPIO.setup(b_pin, GPIO.IN)
        GPIO.setup(a_pin, GPIO.OUT)
        count = 0
        GPIO.output(a_pin, True)
        while not GPIO.input(b_pin):
            count = count + 1
        return count

    def analog_read():
        discharge()
        return charge_time()

    while True:
        print(analog_read())
        time.sleep(1)
```

When you run the program, you should see some output like this:

```
$ sudo python pot_step.py
10
12
10
10
16
23
43
53
67
72
86
105
123
143
170
```

The reading will vary between about 10 and about 170 as you rotate the knob of the trimpot.

Discussion

To explain how this program works, I first need to explain how the *step response* technique can be used to measure the resistance of the variable resistor.

Figure 12-2 shows the schematic diagram for the recipe.

This way of doing things is called *step response* because it works by seeing how the circuit responds from the *step* change when an output is switched from low to high.

You can think of a capacitor as a tank of electricity, and as it fills with charge, the voltage across it increases. You can't measure that voltage directly, because the Raspberry Pi doesn't have an ADC converter. However, you can time how long it takes for the capacitor to fill with charge to the extent that it gets above the 1.65V or so that constitutes

Figure 12-2. Measuring resistance using step response

a high digital input. The speed at which the capacitor fills with charge depends on the value of the variable resistor (Rt). The lower the resistance, the faster the capacitor fills with charge and the voltage rises.

To be able to get a good reading, you must also be able to empty the capacitor each time before you take a reading. In Figure 12-2, connection A is used to charge the capacitor through Rc and Rt, and connection B is used to discharge (empty) the capacitor through Rd. The resistors Rc and Rd are used to make sure that there is no way too much current can flow as the capacitor is charged and discharged.

The steps involved in taking a reading are first to discharge the capacitor through Rd and then to let it charge through Rc and Rt.

To discharge it, connection A (GPIO 18) is set to be an input, effectively disconnecting Rc and Rt from the circuit. Connection B (GPIO 23) is then set to be an output and low. This is held there for 5 milliseconds to empty the capacitor.

Now that the capacitor is empty, you can start to allow charge to flow into it by setting connection B to be an input (effectively disconnecting it) and then enabling connection A to be a high output at 3.3V. Capacitor C will now begin to charge through Rc and Rt. The while loop will then simply count as fast as possible until the voltage at connection B switches from being low to high at about 1.65V.

At that point, the counting stops and the count value is returned.

Figure 12-3 shows how a resistor and capacitor in this kind of arrangement charge and discharge as the voltage is toggled between high and low.

You can see that the voltage at the capacitor increases rapidly at first but then tails off as the capacitor becomes full. Fortunately, you are interested in the area of the curve up

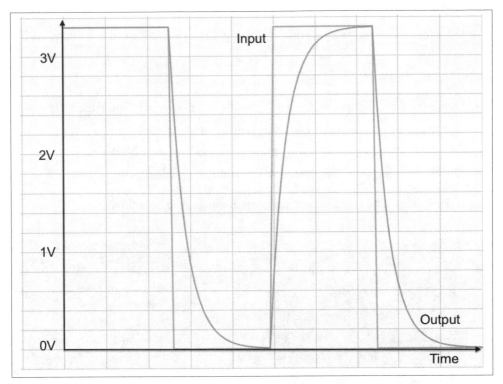

Figure 12-3. Charging and discharging a capacitor

until the capacitor reaches about 1.65V, which is a fairly straight line, meaning that the time taken for the voltage across the capacitor to rise to this point is roughly proportional to the resistance of Rt and hence the position of the knob.

This approach is not hugely accurate, but it is very low cost and easy to use.

See Also

Using a step response works well with all kinds of resistive sensors for light (Recipe 12.2) and even gas detection (Recipe 12.3).

For more accurate measurements of the trimpot position, see Recipe 12.4, where the pot is used with an ADC converter.

12.2. Measuring Light

Problem

You want to measure light intensity with a Raspberry Pi and a photoresistor.

Solution

Use the same basic recipe and code as Recipe 12.1, but replace the trimpot with a photoresistor.

To make this recipe, you will need:

- Breadboard and jumper wires (see "Prototyping Equipment" on page 380)
- Photoresistor (see "Resistors and Capacitors" on page 380)
- Two 1kΩ resistors (see "Resistors and Capacitors" on page 380)
- 220 nF capacitor (see "Resistors and Capacitors" on page 380)

Figure 12-4 shows the arrangement of components on the breadboard.

Figure 12-4. Measuring light on a Raspberry Pi

Using the same program as Recipe 12.1 (*pot_step.py*), you will see the output vary as you move your hand over the photoresistor to cut out some of the light.

Discussion

A photoresistor is a resistor whose resistance varies depending on the amount of light coming through its transparent window. The brighter the light, the lower the resistance. Typically, the resistance varies between about 1kΩ in bright light up to perhaps 100kΩ in complete darkness.

The sensors can only really give a rough idea of the light level.

See Also

You could also use an ADC with the photoresistor (Recipe 12.4).

12.3. Detecting Methane

Problem

You want to measure gas levels using a methane sensor.

Solution

Low-cost resistive gas sensors are available that can easily be wired to a Raspberry Pi to detect gases such as methane. You can use the *step response* method that you first used in Recipe 12.1.

To make this recipe, you will need:

- Breadboard and jumper wires (see "Prototyping Equipment" on page 380)
- Methane sensor (see "Modules" on page 381)
- Two 1kΩ resistors (see "Resistors and Capacitors" on page 380)
- 220 nF capacitor (see "Resistors and Capacitors" on page 380)

The sensor contains a heating element that requires 5V at up to 150 mA. The Raspberry Pi is capable of providing this as long as its power supply can supply the extra 150 mA.

The sensor module has rather thick legs—too thick to fit into breadboard holes. One way around this is to solder short lengths of solid core wire to each lead (Figure 12-5). Another is to buy SparkFun's gas sensor breakout board (*https://www.sparkfun.com/products/8891*).

Wire the breadboard as shown in Figure 12-6 if you're using the SparkFun breakout board, or as shown in Figure 12-7 if you soldered longer leads to the gas sensor.

Note that the direct connection shown in Figure 12-7 uses the same symbol for the breakout board rather than the sensor on its own, but if you look carefully the connections are to the six sensor pins, not the four pins of the breakout.

Figure 12-5. Soldering leads onto the gas sensor

Figure 12-6. Connecting a methane gas sensor to a Raspberry Pi (breakout board)

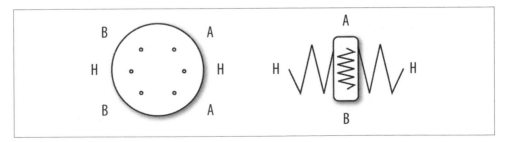

Figure 12-7. Connecting a methane gas sensor to a Raspberry Pi (direct)

You can use the exact same program as Recipe 12.1, and you can test out the methane sensor by breathing on it. You should see the readings from the sensor drop when you breathe on it.

Discussion

The obvious use of a methane gas sensor is for novelty *fart detecting* projects. A more serious use would be for detecting leaks of natural gas. You could, for instance, imagine a Raspberry Pi home watch project that monitored the home with various sensors. It could then send you an email, while you are on vacation, informing you that your house is about to explode. Or maybe not.

These types of sensors (Figure 12-8) use a heating element that warms a resistive surface impregnated with a catalyst sensitive to a particular gas. When the gas is present, the resistance of the catalyst layer changes.

Figure 12-8. A methane gas sensor

Both the heater and the sensing surface are electrically just resistors. So both can be connected either way around.

This particular gas sensor is most sensitive to methane, but will also detect other gases to a lesser extent. That is why breathing on the sensor alters the reading, as healthy individuals will not normally breathe out methane. The cooling effect of blowing on the element may also have an effect.

See Also

The datasheet for this sensor can be found at *http://bit.ly/1gYupsu*. This will give you all sorts of useful information about the sensor's sensitivity to various gases.

There are a range of these low-cost sensors available for sensing different gases. For a list of sensors offered by SparkFun, see the SparkFun website (*https://www.spark fun.com/categories/146*).

12.4. Measuring a Voltage

Problem

You want to measure an analog voltage.

Solution

The Raspberry Pi GPIO connector has only digital inputs. If you wish to measure a voltage, you need to use a separate analog-to-digital converter (ADC).

Use the MCP3008 eight-channel ADC chip. This chip actually has eight analog inputs, so you can connect up to eight sensors to one of these and interface to the chip using the Raspberry Pi SPI interface.

To make this recipe, you will need:

- Breadboard and jumper wires (see "Prototyping Equipment" on page 380)
- MCP3008 eight-channel ADC IC (see "Integrated Circuits" on page 381)
- 10kΩ trimpot (see "Resistors and Capacitors" on page 380)

Figure 12-9 shows the breadboard layout for using this chip. Make sure that you get the chip the right way around. The little notch in the package should be toward the top of the breadboard.

The variable resistor has one end connected to 3.3V and the other to ground, which allows the middle connection to be set to any voltage between 0 and 3.3V.

Figure 12-9. Using a MCP3008 ADC IC with a Raspberry Pi

Before trying the program, make sure you have SPI enabled and the SPI Python library installed (Recipe 8.6).

Open an editor (nano or IDLE) and paste in the following code. As with all the program examples in this book, you can also download the program from the Code section of *http://www.raspberrypicookbook.com*, where it is called *adc_test.py*.

```
import spidev, time

spi = spidev.SpiDev()
spi.open(0, 0)

def analog_read(channel):
    r = spi.xfer2([1, (8 + channel) << 4, 0])
    adc_out = ((r[1]&3) << 8) + r[2]
    return adc_out

while True:
    reading = analog_read(0)
    voltage = reading * 3.3 / 1024
    print("Reading=%d\tVoltage=%f" % (reading, voltage))
    time.sleep(1)
```

The interesting part of the program is contained in the `analog_read` function. This takes a parameter that should be between 0 and 7 and specifies which of the eight analog inputs down the lefthand side of the chip should be read.

The bit manipulation sets up a request for the appropriate channel and then sends the bits to the MCP3008, which reads the resultant data:

```
$ sudo python adc_test.py
Reading=0        Voltage=0.000000
Reading=126         Voltage=0.406055
Reading=221      Voltage=0.712207
Reading=305      Voltage=0.982910
Reading=431      Voltage=1.388965
Reading=527         Voltage=1.698340
Reading=724      Voltage=2.333203
Reading=927      Voltage=2.987402
Reading=10       Voltage=3.296777
Reading=1020     Voltage=3.287109
Reading=1022     Voltage=3.293555
```

Discussion

The MCP3008 has 10-bit ADCs, so when you take a reading, it gives you a number between 0 and 1023. The test program converts this into a voltage reading by multiplying the reading by the voltage range (3.3V) and then dividing it by 1,024.

You can combine any of the following recipes that use the MCP3008 to allow readings to be taken from up to eight sensors.

You can also use resistive sensors with the MCP3008 by combining them with a fixed-value resistor and arranging them as a voltage divider (see Recipe 12.6 and Figure 12-11).

See Also

If you're just interested in detecting the turning of a knob, you can use a rotary encoder instead of a pot (Recipe 11.7).

You can also detect the position of a pot without the use of an ADC chip using the step response method (Recipe 12.1).

Check out the datasheet for the MCP3008 (*http://bit.ly/I1lSqn*).

12.5. Reducing Voltages for Measurement

Problem

You want to measure a voltage, but it is higher than the 3.3V possible using an MCP3008 (Recipe 12.4).

Solution

Use a pair of resistors to act as a voltage divider to reduce the voltage to a suitable range.

To try this recipe, you will need:

- Breadboard and jumper wires (see "Prototyping Equipment" on page 380)
- MCP3008 eight-channel ADC IC (see "Integrated Circuits" on page 381)
- 10kΩ resistor (see "Resistors and Capacitors" on page 380)
- 3.3kΩ resistor (see "Resistors and Capacitors" on page 380)
- 9V battery and clip lead

Figure 12-10 shows the arrangement for this, using a breadboard. The setup will measure the voltage of the battery.

Figure 12-10. Reducing the voltage of analog inputs

 Never use this recipe to measure high-voltage AC, or for that matter, any type of AC. It is for low-voltage DC only.

Open an editor (nano or IDLE) and paste in the following code. As with all the program examples in this book, you can also download the program from the Code section of the *Raspberry Pi Cookbook* website (*http://www.raspberrypicookbook.com*), where it is called *adc_scaled.py*.

```
import spidev

R1 = 10000.0
R2 = 3300.0

spi = spidev.SpiDev()
spi.open(0,0)

def analog_read(channel):
```

```
    r = spi.xfer2([1, (8 + channel) << 4, 0])
    adc_out = ((r[1]&3) << 8) + r[2]
    return adc_out

reading = analog_read(0)
voltage_adc = reading * 3.3 / 1024
voltage_actual =  voltage_adc / (R2 / (R1 + R2))
print("Battery Voltage=" + str(voltage_actual))
```

The program is very similar to that of Recipe 12.4. The main difference is the scaling, using the values of the two resistors. The values of these two resistors are held in the variables *R1* and *R2*.

When you run the program, the battery voltage will be displayed:

```
$ sudo python adc_scaled.py
Battery Voltage=8.62421875
```

 Read the discussion carefully before attaching anything of higher than 9V, or you may destroy the MCP3008.

Discussion

This arrangement of resistors is called a *voltage divider* or sometimes a *potential divider* (Figure 12-11). The formula for calculating the output voltage, given the input voltage and the values of the two resistors, is:

Vout = Vin * R2 / (R1 + R2)

Figure 12-11. A voltage divider

This means that if R1 and R2 were both the same value (say, 1kΩ) then *Vout* would be half of *Vin*.

When choosing R1 and R2, you also need to consider the current flowing through R1 and R2. This will be Vin/(R1 + R2). In the preceding example, R1 is 10kΩ and R2 is 3.3kΩ. So the current flowing will be 9V/13.3kΩ = 0.68 mA. This is low, but still enough to eventually drain the battery, so do not leave it connected all the time.

See Also

To avoid the math, you can use an online resistor calculator (*http://bit.ly/17EtrJI*).

The voltage divider is also used to convert resistance to voltage when using a resistive sensor with an ADC (Recipe 12.6).

12.6. Using Resistive Sensors with an ADC

Problem

You have a resistive sensor that you wish to use with an MCP3008 ADC chip.

Solution

Use a potential divider with one fixed resistor and the resistive sensor to convert the resistance of the sensor into a voltage that can be measured with the ADC.

As an example, you can remake the light sensing Recipe 12.2 to use the MCP3008 instead of the step response technique.

To try this recipe, you will need:

- Breadboard and jumper wires (see "Prototyping Equipment" on page 380)
- MCP3008 eight-channel ADC IC (see "Integrated Circuits" on page 381)
- 10kΩ resistor (see "Resistors and Capacitors" on page 380)
- Photoresistor (see "Resistors and Capacitors" on page 380)

Figure 12-12 shows the arrangement for this, using a breadboard.

You can use the exact same program as Recipe 12.4 (*adc_test.py*). Covering up the light sensor with your hand changes the readings. You also need to set up SPI on your Raspberry Pi, so if you haven't already done so, follow Recipe 8.6.

```
$ sudo python adc_test.py

Reading=341      Voltage=1.098926
Reading=342      Voltage=1.102148
Reading=227      Voltage=0.731543
Reading=81       Voltage=0.261035
Reading=86       Voltage=0.277148
```

These readings may be quite a bit different, depending on your photoresistor, but the important thing is that the figure changes as the light level changes.

Figure 12-12. Using a photoresistor with an ADC

Discussion

Choice of fixed value is not very critical. If the value is too high or low, then you will just find that the range of readings is rather narrow. Select a resistor value somewhere between the minimum and maximum resistance of the sensor. You may need to experiment with a few resistors before deciding on one that suits your sensor over the range of readings that you're interested in. If in doubt, start with 10kΩ and see how that works.

You can swap out the photoresistor for pretty much any resistive sensor. So, for instance, you could use the gas sensor of Recipe 12.3.

See Also

To measure light intensity without the complication of an ADC, see Recipe 12.2. For an example of using more than one channel of the ADC at a time, see Recipe 12.8.

12.7. Measuring Temperature with an ADC

Problem

You want to measure temperature using a TMP36 and an analog-to-digital converter.

Solution

Use an MCP3008 ADC chip.

However, unless you need more than one analog channel, you should consider using the DS18B20 digital temperature sensor, which is more accurate and doesn't require a separate ADC chip (Recipe 12.9).

To try this recipe, you will need:

- Breadboard and jumper wires (see "Prototyping Equipment" on page 380)
- MCP3008 eight-channel ADC IC (see "Integrated Circuits" on page 381)
- TMP36 temperature sensor (see "Integrated Circuits" on page 381)

Figure 12-13 shows the arrangement for this, using a breadboard.

Figure 12-13. Using a TMP36 with an ADC

Make sure that you get the TMP36 the right way around. One side of the package is flat, while the other is curved.

You will need to set up SPI on your Raspberry Pi, so if you haven't already done so, follow Recipe 8.6.

Open an editor (nano or IDLE) and paste in the following code. As with all the program examples in this book, you can also download the program from the Code section of the *Raspberry Pi Cookbook* website (*http://www.raspberrypicookbook.com*), where it is called *adc_tmp36.py*.

```
import spidev, time

spi = spidev.SpiDev()
spi.open(0,0)

def analog_read(channel):
    r = spi.xfer2([1, (8 + channel) << 4, 0])
    adc_out = ((r[1]&3) << 8) + r[2]
    return adc_out

while True:
    reading = analog_read(0)
    voltage = reading * 3.3 / 1024
    temp_c = voltage * 100 - 50
    temp_f = temp_c * 9.0 / 5.0 + 32
    print("Temp C=%f\t\tTemp f=%f" % (temp_c, temp_f))
    time.sleep(1)
```

The program is based on that of Recipe 12.4. A little bit of additional math calculates the temperature in degrees Celsius and Fahrenheit:

```
$ sudo python adc_tmp36.py
Temp C=19.287109        Temp f=66.716797
Temp C=18.642578        Temp f=65.556641
Temp C=18.964844        Temp f=66.136719
Temp C=20.253906        Temp f=68.457031
Temp C=20.898438        Temp f=69.617188
Temp C=20.576172        Temp f=69.037109
Temp C=21.865234        Temp f=71.357422
Temp C=23.154297        Temp f=73.677734
Temp C=23.476562        Temp f=74.257812
Temp C=23.476562        Temp f=74.257812
Temp C=24.121094        Temp f=75.417969
Temp C=24.443359        Temp f=75.998047
Temp C=25.087891        Temp f=77.158203
```

Discussion

The TMP36 outputs a voltage that is proportional to the temperature. According to the datasheet for the TMP36, the temperature in degrees C is calculated as the voltage (in volts) times 100 minus 50.

The TMP36 is fine for measuring the approximate temperature but is specified as having an accuracy of only 2%. This will only get worse if you attach long leads to it. To some extent, you can calibrate an individual device, but for better accuracy, use a DS18B20 (Recipe 12.9), which has a stated accuracy of 0.5% over a temperature range of -10 to +85 degrees C. Being a digital device, it should not suffer any loss of accuracy when attached to long leads.

See Also

Take a look at the TMP36 datasheet (*http://bit.ly/1b3MfIm*).

12.8. Measuring Acceleration

Problem

You want to connect a triple-axis accelerometer to a Raspberry Pi.

Solution

Use an analog accelerometer with a MCP3008 ADC chip to measure the X, Y, and Z analog outputs.

To try this recipe, you will need:

- Breadboard and jumper wires (see "Prototyping Equipment" on page 380)
- MCP3008 eight-channel ADC IC (see "Integrated Circuits" on page 381)
- ADXL335 triple-axis accelerometer (see "Modules" on page 381)

Figure 12-14 shows the arrangement for this, using a breadboard. It uses three channels of the ADC to measure the X, Y, and Z acceleration forces.

Figure 12-14. Using a triple-axis accelerometer

You will need to set up SPI on your Raspberry Pi, so if you have not already done so, follow Recipe 8.6.

Open an editor (nano or IDLE) and paste in the following code. As with all the program examples in this book, you can also download the program from the Code section of the *Raspberry Pi Cookbook* website (*http://www.raspberrypicookbook.com*), where it is called *adc_accelerometer.py*.

```
import spidev, time

spi = spidev.SpiDev()
spi.open(0,0)

def analog_read(channel):
    r = spi.xfer2([1, (8 + channel) << 4, 0])
    adc_out = ((r[1]&3) << 8) + r[2]
    return adc_out

while True:
    x = analog_read(0)
    y = analog_read(1)
    z = analog_read(2)
    print("X=%d\tY=%d\tZ=%d" % (x, y, z))
    time.sleep(1)
```

The program simply reads the three forces and prints them out:

```
$ sudo python adc_accelerometer.py
X=508    Y=503    Z=626
X=508    Y=504    Z=624
X=506    Y=505    Z=627
X=423    Y=517    Z=579
X=411    Y=513    Z=548
X=532    Y=510    Z=623
X=609    Y=518    Z=495
X=607    Y=521    Z=496
X=610    Y=513    Z=499
```

The first three readings took place with the accelerometer level. In the next three, the whole breadboard was tipped to one side. You can see that the X reading has decreased. Tipping the breadboard the other way causes the X reading to increase.

Discussion

The most common use for an accelerometer is to detect tilt. This works because the Z axis force is dominated by the pull of gravity (Figure 12-15).

When the accelerometer is tilted in one direction, some of that vertical force of gravity becomes active on another axis of the accelerometer.

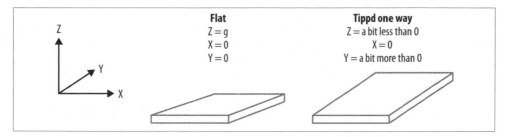

Figure 12-15. Detecting tilt with an accelerometer

We can use this principle to detect when the tilt is past a certain threshold. The following program (*tilt.py*) illustrates this point:

```python
import spidev, time

spi = spidev.SpiDev()
spi.open(0,0)

def analog_read(channel):
    r = spi.xfer2([1, (8 + channel) << 4, 0])
    adc_out = ((r[1]&3) << 8) + r[2]
    return adc_out

while True:
    x = analog_read(0)
    y = analog_read(1)
    z = analog_read(2)
    if x < 450:
        print("Left")
    elif x > 550:
        print("Right")
    elif y < 450:
        print("Back")
    elif y > 550:
        print("Forward")
    time.sleep(0.2)
```

When you run the program, you will start to see direction messages. You could use this to control a roving robot or a motorized pan-tilt head with a webcam attached:

```
$ sudo python tilt.py
Left
Left
Right
Forward
Forward
Back
Back
```

See Also

See the datasheet for the accelerometer chip used in the module (*http://bit.ly/19zQzZq*).

There are many other analog accelerometer modules available. You may find that they give different readings. Make sure that the analog outputs do not exceed 3.3V.

12.9. Measuring Temperature Using a Digital Sensor

Problem

You want to measure temperature using an accurate digital sensor.

Solution

Use the DS18B20 digital temperature sensor. This device is more accurate than the TMP36 used in Recipe 12.7 and uses a one-wire digital interface, so it doesn't require an ADC chip.

Although called *one-wire*, this just refers to the data pin. You do need at least one other wire to connect to a one-wire device.

To make this recipe, you will need:

- Breadboard and jumper wires (see "Prototyping Equipment" on page 380)
- DS18B20 temperature sensor (see "Integrated Circuits" on page 381)
- 4.7kΩ resistor (see "Resistors and Capacitors" on page 380)

Fit the components onto the breadboard as shown in Figure 12-16. Make sure that you get the DS18B20 the right way around.

Both Occidentalis and newer versions of Raspbian have support for the one-wire interface used by the DS18B20 enabled. If the upcoming program doesn't work, then try:

```
$ sudo apt-get upgrade
```

Open an editor (nano or IDLE) and paste in the following code. As with all the program examples in this book, you can also download the program from the Code section of the *Raspberry Pi Cookbook* website (*http://www.raspberrypicookbook.com*), where it is called *temp_DS18B20.py*.

```
import os, glob, time

os.system('modprobe w1-gpio')
os.system('modprobe w1-therm')

base_dir = '/sys/bus/w1/devices/'
device_folder = glob.glob(base_dir + '28*')[0]
```

Figure 12-16. Connecting a DS18B20 to a Raspberry Pi

```
device_file = device_folder + '/w1_slave'

def read_temp_raw():
    f = open(device_file, 'r')
    lines = f.readlines()
    f.close()
    return lines

def read_temp():
    lines = read_temp_raw()
    while lines[0].strip()[-3:] != 'YES':
        time.sleep(0.2)
        lines = read_temp_raw()
    equals_pos = lines[1].find('t=')
    if equals_pos != -1:
        temp_string = lines[1][equals_pos+2:]
        temp_c = float(temp_string) / 1000.0
        temp_f = temp_c * 9.0 / 5.0 + 32.0
        return temp_c, temp_f

while True:
        print("temp C=%f\ttemp F=%f" % read_temp())
        time.sleep(1)
```

When the program is run, it will report the temperature once a second in both degrees Celsius and Fahrenheit:

```
$ python temp_DS18B20.py
temp C=25.187000    temp F=77.336600
temp C=25.125000        temp F=77.225000
temp C=25.062000        temp F=77.111600
```

```
temp C=26.312000          temp F=79.361600
temp C=27.875000          temp F=82.175000
temp C=28.875000          temp F=83.975000
```

Discussion

At first sight, the program looks a little odd. The interface to the DS18B20 uses a file-like interface. The file interface for the device will always be in the folder */sys/bus/w1/devices/* and the name of the file path will start with *28*, but the rest of the file path will be different for each sensor. The code assumes that there will only be one sensor and finds the first folder starting with *28*. Within that folder will be a file called *w1_slave*, which is opened and read to find the temperature.

The sensor actually returns strings of text like this:

```
81 01 4b 46 7f ff 0f 10 71 : crc=71 YES
81 01 4b 46 7f ff 0f 10 71 t=24062
```

The remainder of the code extracts the temperature part of this message. This appears after t= and is the temperature in one-thousandths of a degree Celsius.

The read_temp function calculates the temperature in degrees Fahrenheit and returns both.

Besides the basic chip version of the DS18B20, you can also buy a version encapsulated in a rugged and waterproof probe.

See Also

To find out about logging readings, see Recipe 12.12.

This recipe is heavily based on an Adafruit tutorial (*http://bit.ly/1ioBYbh*).

To measure temperature using a less accurate TMP36 analog sensor, see Recipe 12.7.

Take a look at the datasheet for the DS18B20 (*http://bit.ly/17EveOI*).

12.10. Measuring Distance

Problem

You want to measure distance using an ultrasonic rangefinder.

Solution

Use a low cost SR-04 rangefinder. These devices need two GPIO pins, one to trigger the pulse of ultrasound and the other to monitor how long it takes for the echo to return.

To make this recipe, you will need:

- Breadboard and jumper wires (see "Prototyping Equipment" on page 380)
- SR-04 rangefinder (eBay)
- 470Ω resistor (see "Resistors and Capacitors" on page 380)
- 270Ω resistor (see "Resistors and Capacitors" on page 380)

Fit the components onto the breadboard as shown in Figure 12-17. The resistors are necessary to reduce the *echo* output of the rangefinder from 5V to 3.3V (see Recipe 8.12).

Figure 12-17. Connecting a SR-04 rangefinder to a Raspberry Pi

Open an editor (nano or IDLE) and paste in the following code. As with all the program examples in this book, you can also download the program from the Code section of the *Raspberry Pi Cookbook* website (*http://www.raspberrypicookbook.com*), where it is called *ranger.py*.

```python
import RPi.GPIO as GPIO
import time

trigger_pin = 18
echo_pin = 23

GPIO.setmode(GPIO.BCM)
GPIO.setup(trigger_pin, GPIO.OUT)
GPIO.setup(echo_pin, GPIO.IN)

def send_trigger_pulse():
    GPIO.output(trigger_pin, True)
    time.sleep(0.0001)
    GPIO.output(trigger_pin, False)

def wait_for_echo(value, timeout):
    count = timeout
    while GPIO.input(echo_pin) != value and count > 0:
```

```
        count = count - 1

    def get_distance():
        send_trigger_pulse()
        wait_for_echo(True, 10000)
        start = time.time()
        wait_for_echo(False, 10000)
        finish = time.time()
        pulse_len = finish - start
        distance_cm = pulse_len / 0.000058
        distance_in = distance_cm / 2.5
        return (distance_cm, distance_in)

    while True:
        print("cm=%f\tinches=%f" % get_distance())
        time.sleep(1)
```

The operation of the program is described in the Discussion. When the program is run, it will report the distance in both centimeters and inches, once per second. Use your hand or some other obstacle to change the reading.

```
sudo python ranger.py
cm=154.741879    inches=61.896752
cm=155.670889    inches=62.268356
cm=154.865199    inches=61.946080
cm=12.948595     inches=5.179438
cm=14.087249     inches=5.634900
cm=13.741954     inches=5.496781
cm=20.775302     inches=8.310121
cm=20.224473     inches=8.089789
```

Discussion

While there are a number of ultrasonic rangefinders available, the type used here is easy to use and low cost. It works by sending a pulse of ultrasound and then measuring the amount of time taken for the echo to be received. One of the round ultrasonic transducers on the front of the device is the transmitter, and the other is the receiver.

This process is controlled from the Raspberry Pi. The difference between this type of device and more expensive models is that the more expensive versions include their own microcontroller, which carries out all the necessary timing and provides an I2C or serial interface to return a final reading.

When you are using one of these sensors with a Raspberry Pi, the *trig* (trigger) input to the rangefinder is connected to a GPIO output, and the *echo* output of the rangefinder is connected to a GPIO input on the Raspberry Pi after having its voltage range lowered from 5V to a safe 3.3V.

Figure 12-18. The oscilloscope trace for trigger and echo

Figure 12-18 shows an oscilloscope trace of the sensor in action. The top (red) trace is connected to *trig* and the bottom (yellow) trace is connected to *echo*. You can see that first the *trig* pin is taken high for a short pulse. There is then a short delay before the *echo* pin goes high. This then stays high for a period that is proportional to the distance from the sensor.

The code for this sensor first generates a trigger pulse (using the function `send_trig ger_pulse`). It must wait until the *echo* pin goes high and then times how long the *echo* pin stays high.

We can then calculate the distance using the time taken for the echo pulse and the speed of sound.

The `wait_for_echo` function waits until the *echo* pin either goes high or low, depending on its first argument. Its second argument is used to provide a timeout so that if for any reason, the *echo* pin does not change to the state being waited for, the loop won't hang indefinitely.

See Also

Take a look at the datasheet for the ultrasonic rangefinder (*http://bit.ly/183GLgz*).

Figure 12-19. Displaying a sensor reading using Tkinter

12.11. Displaying Sensor Values

Problem

You have a sensor wired to your Raspberry Pi, but want a big digital display of the reading on the screen.

Solution

Use the Tkinter library to open a window and write the reading on it as a label with a large font (Figure 12-19).

This example uses data from the ultrasonic rangefinder of Recipe 12.10. So, complete that recipe first if you want to try out this example.

To test out the rangefinder, open an editor (nano or idle) and paste in the following code. As with all the program examples in this book, you can also download the program from the code section of the book's website (*http://www.raspberrypicookbook.com*), where it is called *gui_sensor_reading.py*. The code for measuring distances using the ultrasonic sensor is the same as Recipe 12.10, so only the section of the code concerned with displaying the value is repeated here:

```
class App:

    def __init__(self, master):
        self.master = master
        frame = Frame(master)
        frame.pack()
```

```
        label = Label(frame, text='Distance (inches)', font=("Helvetica", 32))
        label.grid(row=0)
        self.reading_label = Label(frame, text='12.34', font=("Helvetica", 110))
        self.reading_label.grid(row=1)
        self.update_reading()

    def update_reading(self):
        cm, inch = get_distance()
        reading_str = "{:.2f}".format(inch)
        self.reading_label.configure(text=reading_str)
        self.master.after(500, self.update_reading)

root = Tk()
root.wm_title('Range Finder')
app = App(root)
root.geometry("400x300+0+0")
root.mainloop()
```

Discussion

Athough this recipe uses a distance sensor, it would work equally well with the other sensor recipes in this chapter. You just need to change the labels and the method of obtaining a reading from the sensor.

See Also

For information on formatting numbers to a certain number of decimal places, see Recipe 7.1.

12.12. Logging to a USB Flash Drive

Problem

You want to log data measured with a sensor onto a USB flash drive.

Solution

Write a Python program that writes the data to a file on a USB flash drive. By writing the file in CSV (comma-separated values), you can import directly into a spreadsheet, including Gnumeric on the Raspberry Pi (Recipe 4.2).

The example program will log temperature readings recorded from a DS18B20. So, if you wish to try this out, first build Recipe 12.9.

Open an editor (nano or IDLE) and paste in the following code. As with all the program examples in this book, you can also download the program from the Code section of

the *Raspberry Pi Cookbook* website (*http://www.raspberrypicookbook.com*), where it is called *temp_log.py*.

```python
import os, glob, time, datetime

log_period = 600 # seconds

logging_folder = glob.glob('/media/*')[0]
dt = datetime.datetime.now()
file_name = "temp_log_{:%Y_%m_%d}.csv".format(dt)
logging_file = logging_folder + '/' + file_name

os.system('modprobe w1-gpio')
os.system('modprobe w1-therm')

base_dir = '/sys/bus/w1/devices/'
device_folder = glob.glob(base_dir + '28*')[0]
device_file = device_folder + '/w1_slave'

def read_temp_raw():
    f = open(device_file, 'r')
    lines = f.readlines()
    f.close()
    return lines

def read_temp():
    lines = read_temp_raw()
    while lines[0].strip()[-3:] != 'YES':
        time.sleep(0.2)
        lines = read_temp_raw()
    equals_pos = lines[1].find('t=')
    if equals_pos != -1:
        temp_string = lines[1][equals_pos+2:]
        temp_c = float(temp_string) / 1000.0
        temp_f = temp_c * 9.0 / 5.0 + 32.0
        return temp_c, temp_f

def log_temp():
    temp_c, temp_f = read_temp()
    dt = datetime.datetime.now()
    f = open(logging_file, 'a')
    f.write('\n"{:%H:%M:%S}",'.format(dt))
    f.write(str(temp_c))
    f.close()

print("Logging to: " + logging_file)
while True:
    log_temp()
    time.sleep(log_period)
```

The program is set to log the temperature every 10 minutes (600 seconds). You can alter this by changing the value of log_period.

Discussion

When you plug a USB flash drive in to a Raspberry Pi, it automatically installs it under */media*. If there is more than one drive, then the program uses the first folder it finds inside */media*. The name of the logging file is constructed from the current date.

If you open the file in a spreadsheet like Open Office, you will be able to edit it directly. Your spreadsheet may ask you to specify the separator for the data, which will be a comma.

Figure 12-20 shows a set of data captured using this recipe, and the resulting file has been opened with the Gnumeric spreadsheet running on the Raspberry Pi.

Figure 12-20. Charting data

See Also

This program could easily be adapted for use with other types of sensor, such as light (Recipe 12.2) or acceleration (Recipe 12.8).

CHAPTER 13
Displays

13.0. Introduction

Although the Raspberry Pi can use a monitor or TV as a display, often it is nice to use a smaller, more specialized display with it. In this chapter, we explore a range of different displays that can be attached to a Raspberry Pi and learn how to use them.

Most of the recipes require the use of a solderless breadboard and male-to-female jumper wires (see Recipe 8.10).

13.1. Using a Four-Digit LED Display

 Be sure to check out the accompanying video for this recipe at *http:// razzpisampler.oreilly.com*.

Problem

You want to display a four-digit number in an old-fashioned, seven-segment LED display.

Solution

Use an I2C LED module, such as the model shown in Figure 13-1 attached via a breadboard to a Raspberry Pi.

To make this recipe, you need:

- Breadboard and jumper wires (see "Prototyping Equipment" on page 380)
- Adafruit 4 × 7-segment LED with I2C backpack (see "Modules" on page 381)

Figure 13-1. Seven-segment LED display with Raspberry Pi

Figure 13-2 shows the arrangement of components on the breadboard.

For this recipe to work, you will also need to set up your Raspberry Pi for I2C, so follow Recipe 8.4 first.

The display has an accompanying Python library written by Adafruit. It isn't installed as a *proper* library, so to use it, you first need to download the folder structure. If you do not have Git installed, install it now with the following command (see Recipe 3.19).

```
$ sudo apt-get install git
```

Now, you can download the folder structure from GitHub:

```
$ git clone https://github.com/adafruit/Adafruit-Raspberry-Pi-Python-Code.git
```

Change directory into the Adafruit code using:

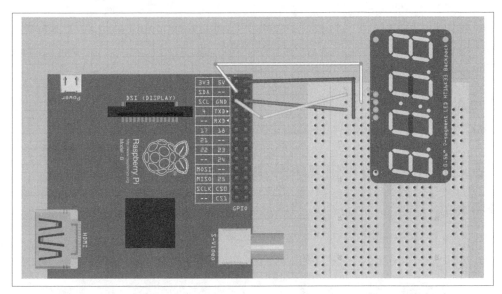

Figure 13-2. The breadboard layout for an LED display with Raspberry Pi

```
$ cd Adafruit-Raspberry-Pi-Python-Code
$ cd Adafruit_LEDBackpack
```

In this folder, you will find a test program that will display the time. Run it using the command:

```
$ sudo python ex_7segment_clock.py
```

Discussion

If you open the example file *ex_7segment_clock.py* in nano, you'll see that the key commands are:

```
from Adafruit_7Segment import SevenSegment
```

which import the library code into your program. You then need to create a instance of SevenSegment using the next line of code. The address supplied as an argument is the I2C address (see Recipe 8.4).

Every I2C slave device has an address number. The LED board has three pairs of solder pads on the back that can be bridged with solder if you want to change the address. This is essential if you need to operate more than one of the displays from a single Raspberry Pi.

```
segment = SevenSegment(address=0x70)
```

To actually set the contents of a particular digit, use a line like this one:

```
segment.writeDigit(0, int(hour / 10))
```

The first argument (0) is the digit position. Note that these positions are 0, 1, 3, and 4. Position 2 is reserved for the two dots in the center of the display.

The second argument is the number to display.

See Also

You can find out more about the Adafruit library at *http://bit.ly/HQBE6W*.

13.2. Displaying Messages on an I2C LED matrix

Problem

You want to control the pixels of a multicolor LED matrix display.

Solution

Use an I2C LED module, such as the model shown in Figure 13-3 attached via a breadboard to a Raspberry Pi. Note that the LED in the breadboard layout looks a little different to the one in the wiring diagram, but their pinouts are the same.

Figure 13-3. LED matrix display with Raspberry Pi

To make this recipe, you need:

- Breadboard and jumper wires (see "Prototyping Equipment" on page 380)

- Adafruit bicolor LED square-pixel matrix with I2C backpack (see "Modules" on page 381)

Figure 13-4 shows the arrangement of components on the breadboard. In fact, if you do not have any buttons or extra components to add, you could just use female-to-female jumper wires and connect the module directly to the Raspberry Pi.

Figure 13-4. The breadboard layout for an LED matrix display with Raspberry Pi

For this recipe to work, you will also need to set up your Raspberry Pi for I2C, so follow Recipe 8.4 first.

The display has an accompanying Python library written by Adafruit. It's not installed as a *proper* library, so to use it, you first need to download the folder structure. If you don't have Git installed, install it now with the following command (see Recipe 3.19):

```
$ sudo apt-get install git
```

Now, you can download the folder structure from GitHub:

```
$ git clone https://github.com/adafruit/Adafruit-Raspberry-Pi-Python-Code.git
```

Change directory into the Adafruit code using:

```
$ cd Adafruit-Raspberry-Pi-Python-Code
$ cd Adafruit_LEDBackpack
```

In this folder, you will find a test program that displays the time as scrolling digits. Run it using the command:

```
$ sudo python ex_8x8_color_pixels.py
```

Discussion

The program cycles through all the colors for each pixel in turn. The code is listed here, with some of the comments and an unnecessary import removed:

```
import time
from Adafruit_8x8 import ColorEightByEight

grid = ColorEightByEight(address=0x70)

iter = 0

# Continually update the 8x8 display one pixel at a time
while(True):
  iter += 1

  for x in range(0, 8):
    for y in range(0, 8):
      grid.setPixel(x, y, iter % 4 )
      time.sleep(0.02)
```

After the import, an instance of `ColorEightByEight` is created.

The address supplied as an argument to the next line is the I2C address (see Recipe 8.4):

```
grid = ColorEightByEight(address=0x70)
```

Every I2C slave device has an address number. The LED board has three pairs of solder pads on the back that can be bridged with solder if you want to change the address. This is essential if you need to operate more than one of the displays from a single Raspberry Pi.

The variable `iter` gets 1 added to it each time through the loop. The command `grid.setPixel` takes *x* and *y* coordinates as its first two parameters. The final parameter is the color to which the pixel will be set. This is a number between 0 and 3 (0 is off, 1 is green, 2 is red, and 3 is orange).

The variable `iter` is used to generate the number between 0 and 3 using the % operator, which is the modulo remainder (i.e., what is left over when you divide `iter` by 4).

See Also

You can find out more about this product at *http://www.adafruit.com/products/902*.

13.3. Using Pi-Lite

Problem

You want to use a Pi-Lite to display text messages on a 9 × 14 LED array.

Solution

Fit a Pi-Lite onto the GPIO port of your Raspberry Pi, and write a Python program to send it messages to display over the serial connection.

The Pi-Lite communicates using the Raspberry Pi's serial port. So, the first thing that you need to do is disable the Raspberry Pi's serial console by following the instructions in Recipe 8.7, and then install PySerial using the instructions in Recipe 8.8.

The Pi-Lite is a powerful board that covers most of the surface of the Raspberry Pi (Figure 13-5).

Figure 13-5. The Pi-Lite with Raspberry Pi

The Pi-Lite (see "Modules" on page 381) has its own processor that does the work of controlling the array of 9 x 14 LEDs. Serial commands can be sent to it and the default behavior is to simply display any text that is sent as horizontally scrolling text.

Power off the Raspberry Pi and attach the Pi-Lite board. When you power the Raspberry Pi back up, the board puts on a little self-test display using the LEDs.

We can try this board out using the Minicom program (Recipe 8.9), which you can install by typing:

```
$ sudo apt-get install minicom
```

Now open a Minicom session with the command:

```
minicom -b 9600 -o -D /dev/ttyAMA0
```

Type in some text, and you will see it flit across the array of LEDs.

Discussion

Sending messages from Python, rather than using Minicom, is also pretty straightforward. You will, however, need to install the Python serial library (Recipe 8.8) using the command:

```
$ sudo apt-get install python-serial
```

Open an editor (nano or IDLE) and paste in the following code. As with all the program examples in this book, you can also download the program from the Code section at *http://www.raspberrypicookbook.com*, where it is called *pi_lite_message.py*.

```
import serial

ser = serial.Serial('/dev/ttyAMA0', 9600)

while True:
    message = raw_input("Enter message: ")
    ser.write(message)
```

The program prompts you for a message and then sends it to the Pi-Lite:

```
$ sudo python pi_lite_message.py
Enter message: Hello
Enter message:
```

Besides displaying scrolling text, you can also control individual pixels of the display and put the display into a bar graph and other modes (see the link to the official guide in the See Also section).

The following example toggles individual pixels on and off (`pi_lite_rain`):

```
import serial
import random

ser = serial.Serial('/dev/ttyAMA0', 9600)

while True:
    col = random.randint(1, 14)
    row = random.randint(1, 9)
    ser.write("$$$P%d,%d,TOGGLE\r" % (col, row))
```

See Also

The official Getting Started guide for the Pi-Lite can be found at *http://bit.ly/1bAq375*.

You can see *Conway's Game of Life* on a Pi-Lite at *http://www.youtube.com/watch?v=bVavjoeHuak*.

13.4. Displaying Messages on an Alphanumeric LCD

Problem

You want to display text on an alphanumeric LCD display.

Solution

Use an HD44780-compatible LCD module and wire it to the GPIO connector.

You will find many low-cost LCD modules, such as the one shown connected to a Raspberry Pi in Figure 13-6. These have 16 pins, although fortunately, not all of them have to be connected to a GPIO pin.

Figure 13-6. A 16×2 LCD display connected to a Raspberry Pi

These modules are available in a number of sizes and are specified by the number of columns and rows of letters they can display. So, for example, the module used in this recipe is described as 16×2, as it can display two rows each of 16 characters. Other common sizes are 8×1, 16×1, 16×2, 20×2, and 20×4.

To make this recipe, you will need:

- Breadboard and jumper wires (see "Prototyping Equipment" on page 380)
- 16×2 HD44780-compatible LCD module (see "Modules" on page 381)
- Row of 16 header pins (see "Miscellaneous" on page 382)

- 10kΩ trimpot (see "Resistors and Capacitors" on page 380)

Figure 13-7 shows the breadboard layout for connecting the display. The LCD modules are not normally supplied with header pins, so you will need to solder these in place.

The trimpot is used to control the contrast of the display. Note that if it looks like the project has not worked, try moving the knob of the trimpot over its entire range. It may well be that the contrast is so off that the display characters are not visible at all.

Figure 13-7. Wiring an LCD display to a Raspberry Pi

The Adafruit Raspberry Pi example code, available from GitHub (*http://bit.ly/1gHetIt*), includes a library for controlling LCD displays that use the HD44780 driver. Before installing it, follow Recipe 8.3 to install the RPi.GPIO library.

The Adafruit library is not installed as a *proper* library, so to use it, you first need to download the folder structure. If you don't have Git installed, install it now with this command (see Recipe 3.19):

```
$ sudo apt-get install git
```

Now, you can download the folder structure from GitHub:

```
$ git clone https://github.com/adafruit/Adafruit-Raspberry-Pi-Python-Code.git
```

Change directory into the Adafruit code using:

```
$ cd Adafruit-Raspberry-Pi-Python-Code
$ cd Adafruit_CharLCD
```

There is a test program here that displays the time and IP address of the Raspberry Pi. But first, if you are using a newer Raspberry Pi model B, revision 2, you will need to edit the file *Adafruit_CharLCD.py*:

```
$ nano Adafruit_CharLCD.py
```

Search for the line:

```
def __init__(self, pin_rs=25, pin_e=24, pins_db=[23, 17, 21, 22], GPIO = None):
```

Replace the number 21 with 27 so that the line looks as follows, and then save and exit the file:

```
def __init__(self, pin_rs=25, pin_e=24, pins_db=[23, 17, 27, 22], GPIO = None):
```

You can now run the example program with the command:

```
$ sudo python Adafruit_CharLCD_IPclock_example.py
```

Discussion

These displays can operate either with a four- or eight-bit data bus and also require three control pins. The pins are connected as per Table 13-1.

Table 13-1. Raspberry Pi and LCD display pin connections

LCD module pin	GPIO pin	Notes
1	GND	0V
2	+5V	5V logic supply
3	No connection	Contrast control voltage
4	25	RS: Register select
5	GND	RW: Read/write (always write)
6	24	EN: Enable
7-10	No connection	Only used in eight-bit mode
11	23	D4: Data line 4
12	17	D5: Data line 5
13	21	D6: Data line 6
14	22	D7: Data line 7
15	+5V	LED backlight
16	GND	LED backlight -

The Adafruit library file *Adafruit_CharLCD.py* is responsible for setting the values on the data pins and then sending them to the display module. It provides the following functions, which you can use in your programs:

`home()`
> Move to top left.

`clear()`
> Clear all text off the display.

`setCursor(column, row)`
> Set the cursor position from where text will be written.

`cursor()`
> Turn on cursor display.

`noCursor()`
> Turn off cursor display (default).

`message(text)`
> Write the text at the current cursor position.

The following minimal example program shows how easy it is to display a simple message by using the library:

```
from Adafruit_CharLCD import Adafruit_CharLCD
from time import sleep

lcd = Adafruit_CharLCD()
lcd.begin(16,2)

i = 0

while Tue:
    lcd.clear()
        lcd.message('Counting: ' + str(i))
        sleep(1)
        i = i + 1
```

These modules are available in many sizes and are specified by the number of columns and rows of letters they can display. So, for example, the module used in this recipe is described as 16×2, as it can display two rows each of 16 characters. Other common sizes are 8×1, 16×1, 16×2, 20×2, and 20×4.

See Also

This recipe is based on the Adafruit tutorial at the Adafruit learning website (*http://bit.ly/1kf4xWC*).

Adafruit also sells a plug-in *shield* with an LCD display attached that is compatible with this recipe.

Arduino and Raspberry Pi

14.0. Introduction

A common alternative to using an interface board, such as a PiFace (Recipe 8.16), is to connect a Raspberry Pi to an Arduino board, as shown in Figure 14-1.

Figure 14-1. An Arduino Uno board

Arduino boards are superficially a little like a Raspberry Pi in that they are small and essentially a computer. However, Arduino boards are very different from the Raspberry Pi in a number of respects:

- They do not have any interface to keyboard, mouse, or screen.
- They have just 2 KB of RAM and 32 KB of flash for storing programs.
- Their processor runs at just 16 MHz compared with the Raspberry Pi's 700 MHz.

This might lead you to wonder why you would use such an apparently feeble board rather than the Raspberry Pi directly.

The answer is that Arduino boards, the most common being the Arduino Uno, are better than the Raspberry Pi at interfacing to external electronics in several ways. For example, Arduino boards have:

- 14 digital inputs/outputs, like the Raspberry Pi's GPIO pins, but each pin can provide up to 40 mA compared with the Raspberry Pi's 3 mA. This enables them to power more devices without the need for extra electronics.
- 6 analog inputs. This makes connecting analog sensors much easier (see Chapter 12).
- 6 PWM outputs. These outputs are hardware-timed and produce a much more accurate PWM signal than can be achieved with the Raspberry Pi, making it a lot better for controlling servo motors.
- A huge range of plug-in shields for everything from motor control to LCD displays of various sorts.

In many ways, using a Raspberry Pi with an Arduino to handle all the low-level stuff is a good combination, playing to the strengths of both boards. This is taken to its extreme with interface boards that have a GPIO connector and also include Arduino-compatible hardware such as the aLaMode (Recipe 14.13).

14.1. Programming an Arduino from Raspberry Pi

 Be sure to check out the accompanying video for this recipe at *http:// razzpisampler.oreilly.com*.

Problem

You want to run the Arduino IDE on a Raspberry Pi so that you can write and upload programs onto an Arduino.

Solution

The Arduino IDE is available for the Raspberry Pi. It is a little bit slow, but usable. Use these commands to install it:

```
$ sudo apt-get update
$ sudo apt-get install arduino
```

At the time of writing, this installs version 1.0.1, which is not the latest version but will suit the Arduino Uno; it will not, however, work for newer boards like the Leonardo and Due. They can still be used with the Raspberry Pi, but you will need some other computer to program them before connecting them to the Raspberry Pi.

After installation, you will find an Electronics group in your Programs menu (Figure 14-2).

The Arduino IDE connects to the Raspberry Pi through its USB cable to program it. This connection also requires that the serial console be disabled. You can follow Recipe 8.7 to do this, but a second option is to run a script created by Kevin Osborn that both disables the serial console and configures the serial ports and Arduino profiles necessary to get things running. This has the advantage that it also sets up the aLaMode board to be ready for use (Recipe 14.13).

To download and run this script, follow these steps:

```
$ wget https://github.com/wyolum/alamode/blob/master/bundles
  /alamode-setup.tar.gz?raw=true -O alamode-setup.tar.gz
$ tar -xvzf alamode-setup.tar.gz
$ cd alamode-setup
$ sudo ./setup
```

If you have not previously disabled your serial console and are relying on the preceding script to do it, then you will need to reboot for this change to take effect.

```
$ sudo reboot
```

You can now connect your Arduino to your Raspberry Pi. From the Tools menu, select Board and set the board type to Arduino Uno. Then, from the Serial Port option, select */dev/ttyACM0*. To upload a test program that will make the LED on the Arduino blink, select the File menu and then click "Examples, Basic," and finally click Blink. Click on the right-arrow on the toolbar to begin the compile and upload process. If all is well, you should see a "Done Uploading" message in the status area at the bottom of the IDE window.

If you find that the device ttyACM0 is not listed even though your Arduino is plugged in, try restarting the Arduino IDE. If that doesn't work, then you may have to reboot your Raspberry Pi. Leave the Arduino connected while you reboot and restart the Arduino IDE.

Figure 14-2. The Arduino IDE running on Raspberry Pi

Discussion

To get the most out of using Arduino with Raspberry Pi, you need to learn a little Arduino programming. You may find the book *Programming Arduino: Getting Started with Sketches* (McGraw-Hill/Tab Books), by yours truly, helpful.

You can, however, make use of an Arduino without needing to write any code on the Arduino side, using a project called PyFirmata. Recipe 14.3 explains how to use PyFirmata.

See Also

The Arduino IDE setup script came from the blog Bald Wisdom (*http://bit.ly/IFZ5zK*).

14.2. Communicating with the Arduino by Using the Serial Monitor

Problem

You want to display messages sent from an Arduino.

Solution

The Arduino IDE includes a feature called the *serial monitor*, which allows you to both send text messages to the Arduino and see messages from the Arduino over the USB cable.

To try this out, you first need to write a very short Arduino program (programs are called *sketches* in the Arduino world). This sketch will just repeat a message, sending it every second.

The Arduino sketch is listed here. As with all the program examples in this book, you can also download the program from *http://www.raspberrypicookbook.com*, where it is contained in the folder *ArduinoHello*. Just follow the link to this book and then click Code.

Use File → New to create a new sketch and paste the following text into it, before uploading it to the Arduino.

```
void setup()
{
  Serial.begin(9600);
}

void loop()
{
  Serial.println("Hello Raspberry Pi");
  delay(1000);
}
```

As soon as the sketch is uploaded onto the Arduino, it will start sending the message "Hello Raspberry Pi" over serial. You won't see this until you open the serial monitor (Figure 14-3) by clicking the icon that looks like a magnifying glass on the right of the toolbar.

Figure 14-3. Viewing messages with the serial monitor

The serial monitor has a drop-down list in the bottom-right corner, where you can select the baud rate (speed of communication). If this isn't already set to 9600, change it to that value.

Discussion

In some of the later recipes in this chapter (Recipes 14.10 and 14.11), you look at writing your own custom code to communicate with Python programs running on the Raspberry Pi so that you don't need to have the Arduino IDE running.

A more generic approach is to use something called PyFirmata, which avoids the need for any programming on the Raspberry Pi. See Recipe 14.3 for details.

See Also

Arduino is quite easy to learn; here are some links to books and online resources to get you started:

- *Programming Arduino: Getting Started with Sketches* (Tab Books) by Simon Monk

- The official Arduino Getting Started Guide (*http://bit.ly/1GAqel*)
- The Adafruit Arduino lesson series (*http://bit.ly/1dTdI1h*)
- *Arduino Cookbook* (O'Reilly) by Michael Margolis

14.3. Setting Up PyFirmata to Control an Arduino from a Raspberry Pi

Problem

You want to use an Arduino as an interface board for your Raspberry Pi.

Solution

Connect the Arduino to a USB socket of the Raspberry Pi so that the computer can communicate and send power to the Arduino.

Next, install the Firmata sketch onto the Arduino and the PyFirmata onto your Raspberry Pi. This entails installing the Arduino IDE, so if you haven't already done so, follow Recipe 14.1.

The Arduino IDE includes Firmata, so all you have to do to install Firmata onto your Arduino board is to upload a sketch. You will find the sketch at File→Examples→Firmata→StandardFirmata.

Once Firmata is installed, the Arduino waits for communication from the Raspberry Pi.

Now you need to install PyFirmata, the other half of the link. This requires the use of the PySerial library, so follow Recipe 8.8 to install this. The PyFirmata library is maintained on GitHub, so you will also need to install Git by using the command `sudo apt-get install git` (Recipe 3.19).

You can now download and install PyFirmata using these commands:

```
$ git clone https://github.com/tino/pyFirmata.git
$ cd pyFirmata
$ sudo python setup.py install
```

You can try out the PyFirmata library from the Python console. Enter the following commands to turn on the built-in LED on Arduino pin 13 (marked with an *L*) and then turn it off again.

```
$ sudo python
Python 2.7.3 (default, Jan 13 2013, 11:20:46)
[GCC 4.6.3] on linux2
Type "help", "copyright", "credits" or "license" for more information.
```

```
>>> import pyfirmata
>>> board = pyfirmata.Arduino('/dev/ttyACM0')
>>> pin13 = board.get_pin('d:13:o')
>>> pin13.write(1)
>>> pin13.write(0)
>>> board.exit()
```

Discussion

The preceding code first imports the PyFirmata library and then makes an instance of Arduino called board, using the USB interface (/dev/ttyACM0) as its parameter. We can then gain a reference to one of the Arduino pins (in this case, 13) and set it to be a digital output. The d is for digital, 13 is the pin number, and o is for output.

To set the output pin high, you just use write(1) and to set it low write(0). You can also use True and False in place of 1 and 0.

Figure 14-4 shows an Arduino with the rows of connections down both sides of the board.

Figure 14-4. I/O pins on an Arduino Uno

The pins at the top of Figure 14-4 marked 0 to 13 can be used as digital inputs or outputs. Some of these pins are also used for other things. Pins 0 and 1 are used as a serial interface, and are in use when the USB port is being used, and pin 13 is attached to the on-board LED marked with an *L*. The digital I/O pins 3, 5, 6, 9, 10, and 11 have a ~ symbol next to them that indicates that they can be used for PWM output (see Figure 14-4).

On the other side of the board, there is one set of connectors that supply power at 5V and 3.3V and six analog inputs marked A0 to A5.

An Arduino Uno on its own uses about 50 mA, which, given that the Raspberry Pi itself is probably using about 10 times that, makes it perfectly possible to power the Arduino from the USB connection of the Raspberry Pi. However, if you start attaching a lot of external electronics to the Arduino, and the current consumption increases, then you may want to power the Arduino from its own power adapter by using the DC barrel socket. This will accept 7V to 12V DC.

The only real downside of using Firmata is that because all instructions have to come from the Raspberry Pi, it doesn't make much use of Arduino's ability to run independently. For advanced projects, you will probably end up writing your own Arduino code that receives instructions from the Raspberry Pi and/or sends messages to the Raspberry Pi, while it gets on with other tasks.

See Also

The following recipes look at using the full range of Arduino pin features with PyFirmata: 14.4, 14.6, 14.7, 14.8, and 14.9.

You can find the official pyFirmata documentation at pyFirmata's GitHub (*http://bit.ly/ Iucv2r*).

14.4. Writing Digital Outputs on an Arduino from a Raspberry Pi

Problem

You want to control Arduino digital outputs from Python on a Raspberry Pi.

Solution

In Recipe 14.3, you flashed the built-in LED (labeled with an *L*) on the Arduino board. Here, you build on this to attach an external LED and write a short Python program to make it blink.

To make this recipe, you need:

- Arduino Uno (see "Modules" on page 381)
- Breadboard and jumper wires (see "Prototyping Equipment" on page 380)
- 270Ω resistor (see "Resistors and Capacitors" on page 380)
- LED (see "Opto-Electronics" on page 381)

Connect the breadboard, holding the components to the Arduino as shown in Figure 14-5.

Figure 14-5. Wiring diagram for Arduino and LED

If you haven't already done so, follow Recipe 14.3 to set up PyFirmata.

The following Python script makes the LED blink at a rate of about 1 Hz. Open an editor (nano or IDLE) and paste in the following code. As with all the program examples in this book, you can also download the program from the Code section of *http://www.rasp berrypicookbook.com*, where it is called *ardu_flash.py*.

```
import pyfirmata
import time

board = pyfirmata.Arduino('/dev/ttyACM0')
led_pin = board.get_pin('d:10:o')

while True:
    led_pin.write(1)
    time.sleep(0.5)
    led_pin.write(0)
    time.sleep(0.5)
```

Discussion

This is very similar to connecting an LED to a Raspberry Pi (Recipe 9.1). However, note that since Arduino outputs can supply a lot more current than Raspberry Pi outputs, you can use a smaller value resistor and make the LED a bit brighter. Arduino outputs are also 5V rather than 3.3V, so the LED will draw about four times the current that it did in Recipe 9.1.

If you want the user interface to control the LED, like you had in Recipe 9.7 (and shown in Figure 14-6), it's pretty straightforward to modify the code. You can find the modified program called *ardu_gui_switch.py*. Remember, this will not work from the SSH command line. You need to have access to the Raspberry Pi's graphical environment so that you can see the user interface.

Figure 14-6. A user interface for turning things on and off

See Also

See Recipe 9.1 for controlling an LED directly from a Raspberry Pi.

14.5. Using PyFirmata with TTL Serial

Problem

You want to use PyFirmata, but over the serial connection (RXD and TXD on the GPIO connector) rather than by USB.

Solution

Use a level converter to connect the RXD pin of the Raspberry Pi to the Tx pin of the Arduino, and the TXD pin of the Raspberry Pi to the Rx pin of the Arduino.

To make this recipe, you need:

- Arduino Uno (see "Modules" on page 381)
- Breadboard and jumper wires (see "Prototyping Equipment" on page 380)
- 270Ω and 470Ω resistors (see "Resistors and Capacitors" on page 380) or a four-way bidirectional level converter (see "Modules" on page 381)

If you are using the level converter module, then connect the breadboard as shown in Figure 14-7.

Figure 14-7. Wiring diagram, using a level converter module for serial communication with an Arduino

If, on the other hand, you are using the pair of resistors, then connect the breadboard as shown in Figure 14-8.

The Arduino Rx input is fine with just 3.3V from the Raspberry Pi TXD pin; however, the 5V coming from the Arduino Tx pin must be dropped to the 3V expected by the Raspberry Pi.

You will need to set up PyFirmata—see Recipe 14.3. The Arduino side of the project remains exactly the same as Recipe 14.4, where USB is used instead of the serial connection. There is one change that you need to make to the Python program running on the Raspberry Pi—change the device name from */dev/ttyACM0* to */dev/ttyAMA0*, as the serial port has a different device name from the USB interface.

The following Python script makes the LED blink at a rate of about 1 Hz. Open an editor (nano or IDLE) and paste in the following code. As with all the program examples in this book, you can also download the program from the Code section of *http://www.rasp berrypicookbook.com*, where it is called *ardu_flash_ser.py*.

```
import pyfirmata
import time

board = pyfirmata.Arduino('/dev/ttyAMA0')
led_pin = board.get_pin('d:13:o')
```

Figure 14-8. Wiring diagram, using a pair of resistors for serial communication with an Arduino

```
while True:
    led_pin.write(1)
    time.sleep(0.5)
    led_pin.write(0)
    time.sleep(0.5)
```

Discussion

The level conversion is necessary because the Raspberry Pi serial port connections, RXD and TXD, operate at 3.3V, whereas the Arduino Uno operates at 5V. While it is OK for the 5V Arduino to use a 3V signal, the reverse is not true and a 5V signal connected to the 3V RXD pin is likely to damage the Raspberry Pi.

See Also

You can also easily adapt the other examples that use PyFirmata (Recipes 14.6 through 14.9) to use a serial rather than USB connection, simply by changing the device name in the Python program.

14.6. Reading Arduino Digital Inputs Using PyFirmata

Problem

You want to read Arduino digital inputs from Python on a Raspberry Pi.

Solution

Use PyFirmata to read a digital input on the Arduino.

To make this recipe, you need:

- Arduino Uno (see "Modules" on page 381)
- Breadboard and jumper wires (see "Prototyping Equipment" on page 380)
- 1kΩ resistor (see "Resistors and Capacitors" on page 380)
- Tactile push switch (see "Miscellaneous" on page 382)

Connect the breadboard, holding the components to the Arduino as shown in Figure 14-9.

If you haven't already done so, follow Recipe 14.3 to set up PyFirmata.

The following Python script prints out a message every time the switch is pressed. It's very similar to the program in Recipe 11.1 (*switch.py*). Open an editor (nano or IDLE) and paste in the following code. As with all the program examples in this book, you can also download the program from the Code section of *http://www.raspberrypicook book.com*, where it is called *ardu_switch.py*. Just follow the link to this book and then click Code.

```
import pyfirmata
import time

board = pyfirmata.Arduino('/dev/ttyACM0')
switch_pin = board.get_pin('d:4:i')
it = pyfirmata.util.Iterator(board)
it.start()
switch_pin.enable_reporting()

while True:
    input_state = switch_pin.read()
    if input_state == False:
        print('Button Pressed')
        time.sleep(0.2)
```

When you run it, nothing will happen for a second or two, while the Firmata sketch starts and establishes communication with the Raspberry Pi. But, once it starts up, each time you press the button, a message will appear.

```
$ sudo python ardu_switch.py
Button Pressed
Button Pressed
Button Pressed
```

Figure 14-9. Wiring diagram for Arduino and push switch

Discussion

PyFirmata uses the concept of an Iterator to monitor the Arduino input pin. The reasons for this are bound up in the implementation of Firmata. This means that you can't simply *read* the value of an Arduino input pin on demand; instead, you have to create a separate Iterator thread that manages the reading of the switch using the commands:

```
it = pyfirmata.util.Iterator(board)
it.start()
```

You then also have to enable reporting for the pin you are interested in using the command:

```
switch_pin.enable_reporting()
```

A side effect of this mechanism is that when you press Ctrl-C to exit the program, it won't exit properly. There is no nice way to kill the `Iterator` thread other than to open another Terminal window or SSH session and kill the process (Recipe 3.24).

If the only Python process running is this program, you can kill it with the command:

```
$ sudo killall python
```

Simply disconnecting the Arduino from the Raspberry Pi, breaking the communication link, will also cause the Python program to exit.

See Also

This is very similar to connecting a switch directly to a Raspberry Pi (Recipe 11.1), and if you just have one switch, there is no real benefit in using an Arduino like this.

14.7. Reading Arduino Analog Inputs Using PyFirmata

Problem

You want to read Arduino analog inputs from Python on a Raspberry Pi.

Solution

Use PyFirmata to read an analog input on the Arduino.

To make this recipe, you need:

- Arduino Uno (see "Modules" on page 381)
- Breadboard and jumper wires (see "Prototyping Equipment" on page 380)
- 10kΩ trimpot (see "Resistors and Capacitors" on page 380)

Connect the breadboard, holding the components to the Arduino as shown in Figure 14-10.

If you haven't already done so, follow Recipe 14.3 to set up PyFirmata.

The following Python script (*ardu_adc.py*) will display both the raw reading from the analog input and the voltage at the analog input. It is very similar to the program in Recipe 12.4 (*adc_test.py*).

Open an editor (nano or IDLE) and paste in the following code. As with all the program examples in this book, you can also download the program from the Code section of *http://www.raspberrypicookbook.com*, where it is called *ardu_adc.py*.

Figure 14-10. Wiring diagram for Arduino and trimpot

```
import pyfirmata
import time

board = pyfirmata.Arduino('/dev/ttyACM0')
analog_pin = board.get_pin('a:0:i')
it = pyfirmata.util.Iterator(board)
it.start()
analog_pin.enable_reporting()

while True:
    reading = analog_pin.read()
    if reading != None:
        voltage = reading * 5.0
        print("Reading-%f\tVoltage-%f" % (reading, voltage))
        time.sleep(1)
```

The analog reading will be a value between 0.0 and 1.0.

```
$ sudo python ardu_adc.py
Reading=0.000000    Voltage=0.000000
Reading=0.165200    Voltage=0.826000
Reading=0.784000    Voltage=3.920000
Reading=1.000000    Voltage=5.000000
```

Discussion

The program is very similar to that of Recipe 14.6. An `Iterator` must be used, and the same problems of stopping the program apply.

The `if` statement is needed, because if the first `read` is made before the actual reading from the analog input has happened, then the `read` will return `None` rather than a number. The `if` statement effectively causes the program to ignore any null readings.

See Also

To use digital inputs, see Recipe 14.6.

14.8. Analog Outputs (PWM) with PyFirmata

Problem

You want to control the brightness of an LED by using PWM through an Arduino.

Solution

Use PyFirmata to send commands to an Arduino to generate a PWM signal on one of its outputs.

To make this recipe, you need:

- Arduino Uno (see "Modules" on page 381)
- Breadboard and jumper wires (see "Prototyping Equipment" on page 380)
- 270Ω resistor (see "Resistors and Capacitors" on page 380)
- LED (see "Opto-Electronics" on page 381)

Connect the breadboard, holding the components to the Arduino as shown in Figure 14-11. This is the same arrangement as in Recipe 14.4.

If you haven't already done so, follow Recipe 14.3 to set up PyFirmata.

The following Python script (*ardu_pwm.py*) will prompt you to enter a value for the PWM power and then set the LED brightness accordingly. It is very similar to the program in Recipe 9.2.

Open an editor (nano or IDLE) and paste in the following code. As with all the program examples in this book, you can also download the program from the Code section of *http://www.raspberrypicookbook.com*, where it is called *ardu_pwm.py*.

```
import pyfirmata
```

Figure 14-11. Wiring diagram for Arduino PWM control of an LED

```
board = pyfirmata.Arduino('/dev/ttyACM0')
led_pin = board.get_pin('d:10:p')

while True:
    duty_s = raw_input("Enter Brightness (0 to 100):")
    duty = int(duty_s)
    led_pin.write(duty / 100.0)
```

With the value entered as 100, the LED should be at full brightness. The brightness decreases as the number decreases.

```
$ sudo python ardu_pwm.py
Enter Brightness (0 to 100):100
Enter Brightness (0 to 100):50
Enter Brightness (0 to 100):10
Enter Brightness (0 to 100):5
Enter Brightness (0 to 100):
```

Discussion

The sketch for this is actually very straightforward. You define the output as PWM output using the command:

```
led_pin = board.get_pin('d:10:p')
```

The *p* is for PWM. But remember, this only works on Arduino pins marked with a ~ symbol.

We can also modify the slider control (Figure 14-12) in Recipe 9.8 so that it will operate through PiFirmata. You can download this sketch as *ardu_gui_slider*.

Figure 14-12. A user interface for controlling brightness

See Also

Although an Arduino can deliver 40 mA to an output, roughly 10 times the current available on a Raspberry Pi GPIO pin, it's still not enough to directly drive a motor or high-power LED module. For these, you would need to use the circuit described in Recipe 9.4, modified to use an Arduino output pin rather than a Raspberry Pi GPIO pin.

14.9. Controlling a Servo Using PyFirmata

Problem

You want to control the position of a servo motor by using an Arduino.

Solution

Use PyFirmata to send commands to an Arduino to generate the pulses necessary to control the position of a servo motor.

To make this recipe, you need:

- Arduino Uno (see "Modules" on page 381)
- Breadboard and jumper wires (see "Prototyping Equipment" on page 380)
- 1kΩ resistor (see "Resistors and Capacitors" on page 380)
- LED (see "Opto-Electronics" on page 381)

Connect the breadboard as shown in Figure 14-13.

Figure 14-13. Wiring diagram for Arduino controlling a servo motor

If you have not already done so, follow Recipe 14.3 to set up PyFirmata.

The following Python script (*ardu_servo.py*) will prompt you to enter a value for the servo's angle and then set the arm of the servo motor accordingly.

Open an editor (nano or IDLE) and paste in the following code. As with all the program examples in this book, you can also download the program from the Code section of *http://www.raspberrypicookbook.com*, where it is called *ardu_servo.py*.

```
import pyfirmata

board = pyfirmata.Arduino('/dev/ttyACM0')
servo_pin = board.get_pin('d:11:s')

while True:
    angle_s = raw_input("Enter Angle (0 to 180):")
    angle = int(angle_s)
    servo_pin.write(angle)
```

With the value entered as 0, the servo should be at one end of its travel. Changing this to 180 sends it to the other end, and 90 puts it somewhere in the middle.

```
$ sudo python ardu_servo.py
Enter Angle (0 to 180):0
Enter Angle (0 to 180):180
Enter Angle (0 to 180):90
```

Discussion

The sketch for this is actually very straightforward. You define the output as a servo output using the command:

```
led_pin = board.get_pin('d:11:s')
```

The *s* is for servo. This can be used on any of the Arduino digital pins.

If you've built Recipe 10.1, you will notice that, by comparison, there is no jitter of the servo when used with an Arduino in this way.

See Also

The straight Raspberry Pi–only solution to using a servo motor is described in Recipe 10.1.

An alternative to using an Arduino for a large number of servo motors is to use a servo module as described in Recipe 10.2.

14.10. Custom Communication with an Arduino over TTL Serial

Problem

You want to exchange data with an Arduino by using the serial interfaces of both devices, without using PyFirmata.

Solution

Use a level converter or pair of resistors to connect the RXD pin of the Raspberry Pi to the Tx pin of the Arduino, and the TXD pin of the Raspberry Pi to the Rx pin of the Arduino.

To make this recipe, you will need:

- Arduino Uno (see "Modules" on page 381)
- Breadboard and jumper wires (see "Prototyping Equipment" on page 380)

- 270Ω and 470Ω resistors (see "Resistors and Capacitors" on page 380) or a four-way bidirectional level converter (see "Modules" on page 381)
- 10kΩ trimpot (see "Resistors and Capacitors" on page 380)

If you are using the level converter module, then connect the breadboard as shown in Figure 14-14.

Figure 14-14. Wiring diagram for serial communication with an Arduino

If, on the other hand, you are using the pair of resistors, then connect the breadboard as shown in Figure 14-15.

The Arduino Rx input is fine with just 3.3V from the Raspberry Pi TXD pin; however, the 5V coming from the Arduino Tx pin must be dropped to the 3V expected by the Raspberry Pi.

You will need to disable serial port logging and install PySerial by following Recipes 8.7 and 8.8.

The following Python script (*ardu_pi_serial.py*) will prompt you to enter a command of either l or r. If you enter l, then the built-in Arduino LED will toggle on and off. If, on the other hand, you enter r, the Arduino will read the analog value from its analog input A0 and send back a number between 0 and 1023 as the reading.

Open an editor (nano or IDLE) and paste in the following code. As with all the program examples in this book, you can also download the program from the Code section of *http://www.raspberrypicookbook.com*, where it is called *ardu_pi_serial.py*.

Figure 14-15. Wiring diagram for serial communication with an Arduino (resistors)

```python
import serial

ser = serial.Serial('/dev/ttyAMA0', 9600)

while True:
    command = raw_input("Enter command: l - toggle LED, r - read A0 ")
    if command == 'l' :
        ser.write('l')
    elif command == 'r' :
        ser.write('r')
        print(ser.readline())
```

You also need to upload the following sketch, *ArduinoSerial*, onto your Arduino Uno. You can paste it into a new sketch window, and it is available with the rest of this book's programs for download.

```c
#include "SoftwareSerial.h"

int ledPin = 13;
int analogPin = A0;

SoftwareSerial ser(8, 9); // RX, TX

boolean ledOn = false;

void setup()
{
  ser.begin(9600);
  pinMode(ledPin, OUTPUT);
}
```

```
void loop()
{
  if (ser.available())
  {
    char ch = ser.read();
    if (ch == 'l')
    {
      toggleLED();
    }
    if (ch == 'r')
    {
      sendAnalogReading();
    }
  }
}

void toggleLED()
{
  ledOn = ! ledOn;
  digitalWrite(ledPin, ledOn);
}

void sendAnalogReading()
{
  int reading = analogRead(analogPin);
  ser.println(reading);
}
```

Run the program and test that the built-in *L* LED on the Arduino toggles on and off using the l command. Then try setting the trimpot to different positions and take analog readings by sending the r command.

```
$ sudo python ardu_pi_serial.py
Enter command: l - toggle LED, r - read A0 l
Enter command: l - toggle LED, r - read A0 l
Enter command: l - toggle LED, r - read A0 r
0

Enter command: l - toggle LED, r - read A0 r
540

Enter command: l - toggle LED, r - read A0 r
1023
```

Discussion

The code in the Python program uses the PySerial library to open a connection onto the serial port. The main program loop then repeatedly asks for commands and processes them, in both cases by writing the single character command to the serial port.

In the case of the r command, it also reads a line from the serial connection and then prints it.

Note that, despite appearances, here the result of serial.readline is a string. If you need to convert this into a number, you can use the function int to convert it. For example:

```
line = ser.readline()
value = int(line)
```

The Arduino sketch is a little more complicated. The example sketch provided can easily be modified to add more inputs or outputs and respond to more commands.

This sketch does not use the Arduino's hardware serial port, as this is generally used by its USB connection. Instead, an Arduino library called *SoftwareSerial* is used to allow pins 8 and 9 to be used as Rx and Tx, respectively.

As with all Arduino sketches, there must be a setup function that is called one time when the Arduino starts up and a loop function that is called repeatedly.

The setup function begins serial communication and sets the pin for the built-in LED (on Arduino pin 13) to be an output.

The loop function first checks to see if any serial messages have arrived by using the ser.available function call. If there is a message to process, it reads the character, and the following if clauses call the appropriate helper function to process the command.

See Also

For more information about the SoftwareSerial library, see *http://arduino.cc/en/Refer ence/SoftwareSerial*.

The alternative to writing your own custom code for communication is to use PyFirmata (Recipe 14.3).

As well as communicating over serial, you can communicate with the Arduino using I2C (Recipe 14.11).

14.11. Custom Communication with an Arduino over I2C

Problem

You want to exchange data with an Arduino using the I2C interfaces of both devices.

Solution

The I2C bus has the concept of master and slave devices. The master device controls the bus and a number of slave devices can be connected to one master, with each slave having its own address.

You will install a sketch on the Arduino that makes it an I2C slave device and use the *SMBus* Python library to write an I2C program for the Raspberry Pi. The *SDA*, *SCL*, and ground connections of both devices will need to be connected to each other, and the Arduino powered separately or from the 5V supply of the Raspberry Pi.

To make this recipe, you need:

- Arduino Uno (see "Modules" on page 381)
- Male-to-female jumper wires (see "Prototyping Equipment" on page 380)

Connect the breadboard, holding the components to the Arduino as shown in Figure 14-16.

Figure 14-16. Wiring diagram for I2C communication with an Arduino

The Arduino board used here is the latest Arduino Uno R3. If you have an older board that doesn't have the two dedicated SCL and SDA pins, then you can connect the Raspberry Pi SDA and SCL pins to Arduino pins A4 and A5.

Install the following sketch onto the Arduino. You can find it named *ArduinoI2C* in the program downloads for this book at *http://www.raspberrypicookbook.com*.

```
#include <Wire.h>

int SLAVE_ADDRESS = 0x04;
int ledPin = 13;
int analogPin = A0;

boolean ledOn = false;

void setup()
{
    pinMode(ledPin, OUTPUT);
    Wire.begin(SLAVE_ADDRESS);
    Wire.onReceive(processMessage);
    Wire.onRequest(sendAnalogReading);
}

void loop()
{
}

void processMessage(int n)
{
    char ch = Wire.read();
    if (ch == 'l')
    {
      toggleLED();
    }
}

void toggleLED()
{
  ledOn = ! ledOn;
  digitalWrite(ledPin, ledOn);
}

void sendAnalogReading()
{
  int reading = analogRead(analogPin);
  Wire.write(reading >> 2);
}
```

You need to follow Recipe 8.4 to set up Raspberry Pi for I2C communication.

Open an editor (nano or IDLE) and paste in the following code. As with all the program examples in this book, you can also download the program from the Code section of *http://www.raspberrypicookbook.com*, where it is called *ardu_pi_i2c.py*.

```
import smbus
import time

# for RPI revision 1, use "bus = smbus.SMBus(0)"
bus = smbus.SMBus(1)
```

```
# This must match in the Arduino Sketch
SLAVE_ADDRESS = 0x04

def request_reading():
    reading = int(bus.read_byte(SLAVE_ADDRESS))
    print(reading)

while True:
    command = raw_input("Enter command: l - toggle LED, r - read A0 ")
    if command == 'l' :
        bus.write_byte(SLAVE_ADDRESS, ord('l'))
    elif command == 'r' :
        request_reading()
```

The test program is very similar to the one used to demonstrate serial communication in Recipe 14.10. It's designed to show communication in both directions. Entering the l command toggles the Arduino's built-in LED on and off, and using the r command reports a reading from the Arduino's A0 analog input. You can use a male-to-male jumper wire to connect A0 to 3.3V, 5V, or GND on the Arduino to get different readings.

```
$ sudo python ardu_pi_i2c.py
Enter command: l - toggle LED, r - read A0 l
Enter command: l - toggle LED, r - read A0 l
Enter command: l - toggle LED, r - read A0 r
184
Enter command: l - toggle LED, r - read A0
```

Discussion

The Arduino acts as the I2C slave in this arrangement and must therefore be given an *address* so that the master running on the Raspberry Pi can identify it if there is more than one slave device connected. In this case, the address is set to 0x04 and held in the SLAVE_ADDRESS variable.

The setup function in the Arduino then sets up two callback functions that will be used. The function processMessage will be invoked whenever the onReceive event occurs. This will be whenever a command is sent from the Raspberry Pi. The other callback function, sendAnalogReading, is associated with the onRequest event. This occurs when the Raspberry Pi requests data, and will read the analog value, divide it by four to make it fit into a single byte, and then send it back to the Raspberry Pi.

The Python counterpart to this sketch first creates a new instance of SMBus called bus. This takes a single argument of 1. This is the I2C port on the Raspberry Pi to be used. This will be 1 unless you are using the very first version of the Raspberry Pi (revision 1). These boards have a black audio connector rather than the light blue connector of the revision 2 boards. If you have one of these older boards, use 0 as the argument here. The I2C port available at the GPIO connector was swapped between these two revisions.

The program prompts the user for a command (l or r). If the command is l, it writes the character l to the Arduino, which in turn causes the onReceive handler (process Message) to be run. This in turn will call toggleLED.

If, on the other hand, the user enters the r command, then the request_reading function will be called. This will call read_byte in the SMBus library that will cause the onRequest event in the Arduino sketch to be invoked.

You may be wondering why it is OK to connect the 5V Arduino directly to the I2C connections of the Raspberry Pi, without the level converter module that you had to use for serial communications (see Recipe 14.5). The reason is that the I2C bus standard operates at whatever voltage the pull-up resistors connected to SDA and SCL use. In this case, the Arduino Uno has no pull-up resistors connected to the I2C lines. These are provided by the Raspberry Pi, which pulls them up to 3.3V.

 Although the Arduino has no pull-up resistors on the I2C lines, this is not true of all I2C devices. So, if the I2C device operates at 5V, make sure that it has no pull-up resistors (but if it does, you can usually remove them relatively easily).

See Also

For the serial equivalent to this setup, see Recipe 14.10.

14.12. Using Small Arduinos with a Raspberry Pi

Problem

You want to use an Arduino board with a Raspberry Pi, but would like something more compact.

Solution

Use one of the small breadboard-friendly Arduino boards.

Figure 14-17 shows an Arduino Pro Mini board. Boards like this have the great advantage that they can be plugged directly into breadboard along with other components needed for the project. The Pro Mini board is also available in a 3.3V version, which avoids any need for level conversion when you're using it with a Raspberry Pi.

Figure 14-17. Arduino Pro Mini and programmer interface

Discussion

Some of these boards, such as the Pro Mini shown in Figure 14-17, require a USB programming interface. You can program them from the Raspberry Pi or another computer if that proves problematic.

As well as official Arduino boards, you can also find many low-cost clones.

See Also

Other boards to consider are:

- Teensy (*http://www.pjrc.com/teensy/*)
- Arduino Micro (*http://arduino.cc/en/Main/ArduinoBoardMicro*)
- Arduino Nano (*http://arduino.cc/en/Main/ArduinoBoardNano*)

14.13. Getting Started with an aLaMode Board and a Raspberry Pi

Problem

You want to use an aLaMode board to interface external electronics with the Raspberry Pi.

Solution

The aLaMode board (see "Modules" on page 381) shown in Figure 14-18 is a wonderful board that is essentially an Arduino Uno with a Raspberry Pi GPIO socket in one corner. The aLaMode fits neatly onto the Raspberry Pi and allows you to use Arduino shields without the need for lots of extra wires.

Figure 14-18. aLaMode interface board

Note that the board shown in Figure 14-18 is without the header sockets needed to attach Arduino shields.

The installation instructions in Recipe 14.1 will also set up aLaMode in your Arduino environment. From the Tools menu of the Arduino IDE, choose Board Type and then ALaMode (Figure 14-19).

You also need to set the serial port to */devttyS0* before programming the aLaMode, which takes place through the serial connection. This means that you will need to disable the console logging (Recipe 8.7).

Discussion

All the Firmata recipes in this chapter will work with the aLaMode, along with the custom serial communication example of Recipe 14.10. The only modification necessary is in the Python programs communicating with the aLaMode, as they must use the serial port rather than the USB port.

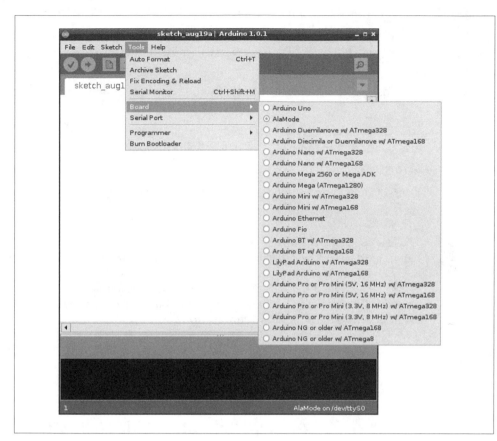

Figure 14-19. aLaMode board type on the Arduino IDE

The aLaMode is hardwired to the Arduino's Tx and Rx pins, in a configuration that is much the same as Recipe 14.5 but with built-in level conversion.

When using Firmata, you need to change the port from */dev/ttyACM0* to */dev/ttyAMA0*:

```
board = pyfirmata.Arduino('/dev/ttyAMA0')
```

Similarly, if you are writing your own custom code, you will need to change the line that opens the serial connection so that it looks like this:

```
ser = serial.Serial('/dev/ttyAMA0', 9600)
```

To try the aLaMode, just load the *Standard Firmata* sketch onto the aLaMode and then run the Python program *ardu_flash_ser.py* (Recipe 14.5).

Some of the interesting features of the aLaMode board are:

- It can be powered from the 5V line of the Raspberry Pi GPIO connector or a separate 5V power adapter connected to its micro-USB socket.

- The RTC is connected directly to the Raspberry Pi rather than the Arduino side of the board.

- Arduino shields can be attached, with a high degree of compatibility (see Recipe 14.14).

- It can be an I2C slave. As well as being linked by serial, the aLaMode's I2C connections are also connected to those of the Raspberry Pi (Recipe 14.11).

- It has a micro SD card reader.

- It has headers for direct connection to servo motors (Recipe 14.9).

 The aLaMode does suffer from one minor design flaw—with the Arduino shield headers soldered in place, the analog inputs A0 to A5 can easily touch the bare metal of the RJ45 Ethernet socket of the Raspberry Pi. To avoid this, place a couple of layers of electrical insulating tape over the top surface of the Ethernet socket on the Raspberry Pi before attaching the aLaMode (Figure 14-20).

See Also

The official aLaMode instructions can be found at *http://bit.ly/1d2YMxh*.

14.14. Using an Arduino Shield with an aLaMode Board and a Raspberry Pi

Problem

You want to use an Arduino shield with a Raspberry Pi.

Solution

Use an aLaMode interface board. You may find that the board requires the Arduino-style header sockets for shields to be soldered into place.

As an example, you'll connect an Arduino LCD shield to the aLaMode (Figure 14-21).

To make this recipe, you need:

- aLaMode Board (see "Modules" on page 381)
- Arduino LCD shield (see "Modules" on page 381)

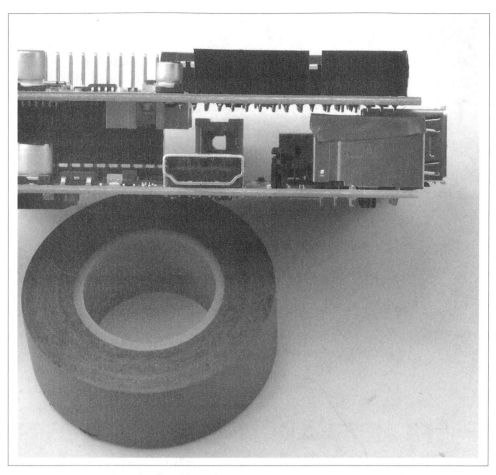

Figure 14-20. Insulating the RJ45 connector on a Raspberry Pi

As with all the program examples in this book, you can also download this sketch from the Code section of *http://www.raspberrypicookbook.com*, where it is called *AlaMode-Shield*.

```
#include <LiquidCrystal.h>

// pins for Freetronics LCD Shield

LiquidCrystal lcd(8, 9, 4, 5, 6, 7);

void setup()
{
  lcd.begin(16, 2);
  lcd.print("Counting!");
}
```

Figure 14-21. Using an LCD Arduino shield with an aLaMode

```
void loop()
{
  lcd.setCursor(0, 1);
  lcd.print(millis() / s1000);
}
```

Simply upload the sketch onto the aLaMode and you should see the message "Counting" appear on the top row of the display, while the bottom row should show a number that counts the seconds.

Discussion

This example used the Freetronics LCD shield. There are a number of other LCD display shields, and some have different pin allocations. If you use a different module, you'll need to change the line:

```
LiquidCrystal lcd(8, 9, 4, 5, 6, 7);
```

The parameters are the Arduino pins to be used for the following LCD connections: rs, enable, d4, d5, d6, and d7, in that order.

See Also

The reference material for the Arduino LiquidCrystal library can be found at *http://arduino.cc/en/Reference/LiquidCrystal*.

14.15. Using Gertboard as an Arduino Interface

Problem

You want to program the ATmega chip on the Gertboard as if it were an Arduino.

Solution

Follow this detailed tutorial (*https://projects.drogon.net/raspberry-pi/gertboard/*) to set up the Arduino IDE, the Raspberry Pi, and the Gertboard to program the Gertboard's processor as if it were an Arduino.

Discussion

The Gertboard ATmega processor operates at 3.3V, removing the need for any level conversion.

See Also

For a basic introduction to the Gertboard, see Recipe 8.17.

The Gertboard manual (*http://bit.ly/17FGFLm*) is available from multiple places.

Parts and Suppliers

Parts

The following tables will help you to find the parts used in this book. Where possible, I have listed product codes for a few suppliers.

There are now many electronic component suppliers that cater to the maker and electronics hobbyist. Some of the most popular are listed in Tables A-1 and A-2.

Table A-1. US suppliers

Supplier	Website	Notes
Adafruit	http://www.adafruit.com	Good for modules
Digikey	http://www.digikey.com/	Wide range of components
MakerShed	http://www.makershed.com/	Good for modules, kits, and tools
MCM Electronics	http://www.mcmelectronics.com/	Wide range of components
Mouser	http://www.mouser.com	Wide range of components
RadioShack	http://www.radioshack.com/	Walk-in stores
SeeedStudio	http://www.seeedstudio.com/	Interesting low-cost modules
SparkFun	http://www.sparkfun.com	Good for modules

Many of the US suppliers also have regional websites and will also ship internationally.

Table A-2. Other suppliers

Supplier	Website	Notes
CPC	http://cpc.farnell.com/	UK-based, wide range of components
Ciseco	http://shop.ciseco.co.uk/	Suppliers of PiLite, Humble Pi, etc.
Farnell	http://www.farnell.com/	International, wide range of components
Maplins	http://www.maplin.co.uk/	UK-based, walk-in shops

Supplier	Website	Notes
Proto-pic	*http://proto-pic.co.uk/*	UK-based, stock SparkFun and Adafruit modules
SK Pang	*http://www.skpang.co.uk/*	UK-based

The other great source of components is eBay.

Searching for components can be time consuming and difficult. The Octopart component search engine (*http://www.octopart.com*) can be very helpful in tracking down parts.

Prototyping Equipment

Many of the hardware projects in this book use jumper wires of various sorts. Male-to-female leads (to connect the Raspberry Pi GPIO connector to a breadboard) and male-to-male (to make connections on the breadboard) are particularly useful. Female-to-female are occasionally useful for connecting modules directly to GPIO pins. You rarely need leads longer than 3 inches (75mm). Table A-3 lists some jumper wire and breadboard specifications, along with their suppliers.

Table A-3. Prototyping equipment

Description	Suppliers
M-M jumper wires	SparkFun: PRT-08431, Adafruit: 759
M-F jumper wires	SparkFun: PRT-09140, Adafruit: 825
F-F jumper wires	SparkFun: PRT-08430, Adafruit: 794
Half-sized breadboard	SparkFun: PRT-09567 Adafruit: 64
Pi Cobbler	Adafruit: 1105

Resistors and Capacitors

Table A-4 lists resistors and capacitors used in this cookbook and some suppliers.

Table A-4. Resistors and capacitors

270Ω 0.25W resistor	Mouser: 293-270-RC
470Ω 0.25W resistor	Mouser: 293-470-RC
1kΩ 0.25W resistor	Mouser: 293-1k-RC
3.3kΩ 0.25W resistor	Mouser: 293-3.3k-RC
4.7kΩ 0.25W resistor	Mouser: 293-4.7k-RC
10 kΩ trimpot	Adafruit: 356, SparkFun: COM-09806, Mouser: 652-3362F-1-103LF
Photoresistor	Adafruit: 161, SparkFun: SEN-09088
220nF capacitor	MCM: 31-0610, Mouser: 80-C322C224M5U5HA

Transistors and Diodes

Table A-5 lists transistors and diodes used in this cookbook and some suppliers.

Table A-5. Transistors and diodes

FQP30N06 N-channel MOSFET transistor	SparkFun: COM-10213, Adafruit: 355
2N3904 NPN bipolar transistor	SparkFun: COM-00521, Adafruit: 756
1N4001 diode	SparkFun: COM-08589, Adafruit: 755

Integrated Circuits

Table A-6 lists integrated circuits used throughout this cookbook and some suppliers.

Table A-6. Integrated circuits

7805 voltage regulator	SparkFun: COM-00107
L293D motor driver	SparkFun: COM-00315, Adafruit: 807
ULN2803 Darlington driver IC	SparkFun: COM-00312, Adafruit: 970
DS18B20 temperature sensor	SparkFun: SEN-00245, Adafruit: 374
MCP3008 eight-channel ADC IC	Adafruit: 856
TMP36 temperature sensor	SparkFun: SEN-10988, Adafruit: 165

Opto-Electronics

Table A-7 lists opto-electronic components used throughout this cookbook and some suppliers.

Table A-7. Opto-electronics

5mm red LED	SparkFun: COM-09590, Adafruit: 299
RGB common cathode LED	SparkFun: COM-11120
TSOP38238 IR sensor	SparkFun: SEN-10266, Adafruit: 157

Modules

Table A-8 lists modules used in this cookbook and some suppliers.

Table A-8. Modules

Raspberry Pi camera module	Adafruit: 1367, MCM: 28-17733, CPC: SC13023
Arduino Uno	SparkFun: DEV-11021, Adafruit: 50, CPC: A000066
Level converter, four-way	SparkFun: BOB-11978, Adafruit: 757
Level converter eight-way	Adafruit: 395
LiPo boost converter/charger	SparkFun: PRT-11231

PowerSwitch tail	Adafruit: 268
16-channel servo controller	Adafruit: 815
Motor driver 1A dual	SparkFun: ROB-09457
RaspiRobot board	Sparkfun: KIT-11561, raspirobot.com
PiFace digital interface board	MCM: 83-14472, CPC: SC12827
Humble Pi	MCM: 83-14637, CPC: SC12871
Pi Plate	Adafruit: 801
Gertboard	MCM: 83-14460, CPC: SC12828
Breakout board with paddle terminals	MCM: 83-14876, CPC: SC12885
PIR motion detector	Adafruit: 189
Venus GPS module	SparkFun: GPS-11058
Methane sensor	SparkFun: SEN-09404
Gas sensor breakout board	SparkFun: BOB-08891
ADXL335 triple-axis accelerometer	Adafruit: 163
4x7-segment LED with I2C backpack	Adafruit: 878
Bicolor LED square-pixel matrix with I2C backpack	Adafruit: 902
PiLite interface board	Ciseco, CPC: SC13018
aLaMode interface board	Makershed: MKWY1, Seeedstudio: ARD10251P
Freetronics Arduino LCD shield	www.freetronics.com
RTC module	Adafruit: 264
16 x 2 HD44780 compatible LCD module	SparkFun: LCD-00255, Adafruit: 181

Miscellaneous

Table A-9 lists miscellaneous tools used in this cookbook and some suppliers.

Table A-9. Miscellaneous

1200mAh LiPo battery	Adafruit: 258
5V relay	SparkFun: COM-00100
5V panel meter	SparkFun: TOL-10285
Servo motor	SparkFun: ROB-09065, Adafruit: 1449
5V 1A power supply	Adafruit: 276
Low power 6V DC motor	Adafruit: 711
0.1 inch header pins	SparkFun: PRT-00116, Adafruit: 392
5 V 5-pin unipolar stepper motor	Adafruit: 858
12 V, 4-pin bipolar stepper motor	Adafruit: 324
Magician chassis with gearmotors	SparkFun: ROB-10825
Tactile push switch	SparkFun: COM-00097, Adafruit: 504

Miniature slide switch	SparkFun: COM-09609, Adafruit: 805
Rotary encoder	Adafruit: 377
4x3 keypad	SparkFun: COM-08653
Piezo buzzer	SparkFun: COM-07950, Adafruit: 160

Index

We'd like to hear your suggestions for improving our indexes. Send email to index@oreilly.com.

About the Author

Dr. Simon Monk (Preston, UK) has a degree in Cybernetics and Computer Science and a PhD in Software Engineering. Simon spent several years as an academic before he returned to the industry, co-founding the mobile software company Momote Ltd. Simon is now a full-time author writing on many aspects of open source hardware, including Raspberry Pi and Arduino as well as a number of general electronics books. You can find out more about his books at http://www.simonmonk.org or follow him on Twitter @simonmonk2.

Colophon

The animal on the cover of *Raspberry Pi Cookbook* is the Eurasian sparrowhawk (*Accipiter nisus*), which also goes by the names northern sparrowhawk or simply sparrowhawk. This small bird of prey is found throughout the Old World. Adult males have bluish gray upper plumage and orange-barred underparts; females and younger birds are all brown with brown-barred underparts. The female is up to 25% larger than the male.

The sparrowhawk specializes in preying on woodland birds, but can be found in any habitat hunting garden birds in towns or cities. Males favor hunting smaller birds—tits, finches, and sparrows; females tend to catch thrushes and starlings, and are capable of killing birds weighing up to 18 ounces (500 grams) or more.

Eurasian sparrowhawks breed in nests that can measure up to 60 cm (2 feet) across, built with twigs. Afterward, four or five pale blue, brown-spotted eggs are laid. Success of breeding relies on females maintaining a high weight; the male delivers food to its mate. After 33 days, the chicks hatch, and they fledge after 24 to 28 days.

A juvenile sparrowhawk will survive its first year at a rate of 34%. After that, its chance of survival more than doubles, with adults surviving from one year to the next at 69% for a typical lifespan of four years. Mortality is greater for young males than for young females. Despite a sharp population decline after WWII, the sparrowhawk is the most common bird of prey in Europe. The use of organochlorine insecticides to treat seeds before sowing has been known to incapacitate or kill sparrowhawks. Those affected lay fragile-shelled eggs that break during incubation. After the chemicals were banned, the sparrowhawk population recovered and remains common, classified as Least Concern by BirdLife International.

The cover image is from *Cassell's Natural History*. The cover fonts are URW Typewriter and Guardian Sans. The text font is Adobe Minion Pro; the heading font is Adobe Myriad Condensed; and the code font is Dalton Maag's Ubuntu Mono.

Have it your way.